Authenticity in the Music of Video Games

Authenticity in the Music of Video Games

Stephanie Lind

LEXINGTON BOOKS
Lanham • Boulder • New York • London

Published by Lexington Books
An imprint of The Rowman & Littlefield Publishing Group, Inc.
4501 Forbes Boulevard, Suite 200, Lanham, Maryland 20706
www.rowman.com

86-90 Paul Street, London EC2A 4NE

Copyright © 2023 by The Rowman & Littlefield Publishing Group, Inc.

All rights reserved. No part of this book may be reproduced in any form or by any electronic or mechanical means, including information storage and retrieval systems, without written permission from the publisher, except by a reviewer who may quote passages in a review.

British Library Cataloguing in Publication Information Available

Library of Congress Cataloging-in-Publication Data

Names: Lind, Stephanie, author.
Title: Authenticity in the music of video games / Stephanie Lind.
Description: Lanham : Lexington Books, 2023. | Includes bibliographical references.
Identifiers: LCCN 2022036863 (print) | LCCN 2022036864 (ebook) | ISBN 9781793627124 (cloth) | ISBN 9781793627148 (paperback) | ISBN 9781793627131 (ebook)
Subjects: LCSH: Video game music—History and criticism. | Video game music—Analysis, appreciation.
Classification: LCC ML3540.7 .L55 2022 (print) | LCC ML3540.7 (ebook) | DDC 781.5/4—dc23/eng/20220802
LC record available at https://lccn.loc.gov/2022036863
LC ebook record available at https://lccn.loc.gov/2022036864

Contents

List of Figures and Tables	vii
Acknowledgments	xiii
Introduction	1
Chapter One: Gameworld	25
Chapter Two: Tropes	59
Chapter Three: Thematic Representation	95
Chapter Four: Retro Gaming Experiences	123
Conclusion	153
Bibliography	171
Index	181
About the Author	193

Figures and Tables

Figure 0.1: Four functions for game sound. Figure by the author. 15

Figure 1.1: Screenshot from the opening bird's-eye-view cut scene, *Assassin's Creed: Odyssey*. Reproduced by permission of Ubisoft Entertainment SA and Hal Leonard, Inc. 29

Figure 1.2: "Poseidon, God of the Sea" with Latin alphabet transliteration. Composed by Giannis Georgantelis. Transcription and analysis by the author. Reproduced by permission of Ubisoft Entertainment SA. 33

Figure 1.3: "Song for a Young Girl" with Latin alphabet transliteration. Composed by Giannis Georgantelis. Transcription and analysis by the author. Reproduced by permission of Ubisoft Entertainment SA and Hal Leonard, Inc. 35

Figure 1.4: "Ares, God of War" with Latin alphabet transliteration. Composed by Giannis Georgantelis. Transcription and analysis by the author. Reproduced by permission of Ubisoft Entertainment SA and Hal Leonard, Inc. 36

Table 1.1: Musical properties of the Greek-language sea shanties in *Assassin's Creed: Odyssey*. Table by the author. 37

Figure 1.5: Tetrachord construction in Ancient Greek scales and variants, with numeric values indicating cents. M. L. West, *Ancient Greek Music* (Oxford: Clarendon Press, 1992), 170. 39

Figure 1.6: "The Painful Way," melody and bass in the first four bars of the song's main theme. Composed by Darren Korb. Reduction and analysis by the author. Original music copyright Supergiant Games. 42

Figure 1.7: "No Escape," melody and bass plus accompaniment rhythm, bars 1–17. Composed by Darren Korb. Reduction and analysis by the author. Original music copyright Supergiant Games. 44

Figure 1.8: Screenshot from the opening cut scene, *Assassin's Creed: Odyssey*. Reproduced by permission of Ubisoft Entertainment SA. 45

Figure 1.9: Analysis, *Assassin's Creed: Odyssey*, Leonidas opening cut scene. Figure by the author. 47

Figure 1.10: Soundscape analysis, opening sequence, *Metro: Exodus*. Figure by the author. 50

Figure 2.1: The instantly recongizable opening of the *Toccata and Fugue in D minor*, BWV 565, by J. S. Bach. Public domain. Analysis by the author. 65

Figure 2.2: "Agro Falls" fragment from "Resurrection," *Shadow of the Colossus*. Composed by Kow Otani. Transcription and analysis by the author. Original music copyright Sony Music Artists, Inc. 68

Figure 2.3: Opening, "A Pane in the Glass," *Stronghold 3*. Composed by Robert L. Euvino. Transcription and analysis by the author. Courtesy Firefly Studios, Inc. 76

Figure 2.4: Plainchant fragments at the beginning of "Amen," *Medieval II: Total War*. Composed by Jeff van Dyck. Plainchant fragments from Catholic Church, *The Parish Book of Chant* (Richmond, VA: Church Music Association of America, 2008). Figure and analysis by the author. 77

Figure 2.5: Plainchant excerpts in "Acre—Underworld," *Assassin's Creed 1*. Composed by Jesper Kyd. Transcription and analysis by the author. Reproduced by permission of Ubisoft Entertainment SA and Hal Leonard, Inc. 77

Figure 2.6: Bars 1–20, "Moldheart's Hornpipe," *Sims Medieval*. Composed by John Debney. Reduction and analysis by the author. Reproduced by permission of Electronic Arts and Hal Leonard, Inc. 80

Figure 2.7: Cadence patterns in "Age of Oppression"/"Age of Aggression," *Skyrim*. Figure by the author. 83

Figure 2.8: Cadences in "Out of the Cold," *Skyrim*. Figure by the author. 84

Figure 2.9: Perfect-fifth leaps in "The Dawn Will Come," *DragonAge: Inquisition*. Composed by Trevor Morris. Reduction and analysis by the author. Courtesy Electronic Arts. 85

Figure 2.10: Screenshot, *Samba de Amigo*. Reproduced by permission of Sega. 87

Table 3.1: Instruments associated with the "Goron City" theme variants throughout the *Legend of Zelda* series. Table by the author. 102

Figure 3.1: Chant melody from *Halo*'s main theme, notated in unmetered rhythm with breakdown into sub-motives. Composed by Martin O'Donnell. Reduction and analysis by the author. Original music copyright Xbox Game Studios. 108

Figure 3.2: Motivic analysis, bars 17–56, "Luck," *Halo 3*. Composed by Martin O'Donnell. Reduction and analysis by the author. Original music copyright Xbox Game Studios. 110

Figure 3.3: Melodic development derived from Motive 3 in "Luck." Boxes highlight pitches in common between the five theme variants. Figure by the author. 111

Figure 3.4: "Keep What You Steal" excerpt, *Halo 3*. Boxes indicate material derived from previous melodic content in "Luck." Composed by Martin O'Donnell. Reduction and analysis by the author. Original music copyright Xbox Game Studios. 113

Figure 3.5: Title Theme with phrase segment analysis, *Sonic the Hedgehog*. Composed by Masato Nakamura. Transcription and analysis by the author. Original music copyright Sega. 115

Figure 3.6: Game Over music with phrase segment analysis, *Sonic the Hedgehog*. Transcription and analysis by the author. Composed by Masato Nakamura. Reproduced by permission of Sega. 116

Figure 3.7: Continue music with phrase segment analysis, *Sonic the Hedgehog*. Transcription and analysis by the author. Composed by Masato Nakamura. Original music copyright Sega. 117

Figure 4.1: Common properties of retro game music. Figure by the author. 128

Figure 4.2: Harmonic analysis, "Soldiers of the Sea," *Return of the Obra Dinn*. Composed by Lucas Pope. Reduction and analysis by the author. Courtesy Lucas Pope. 129

Figure 4.3: Varied accompaniment patterns, "Soldiers of the Sea," *Return of the Obra Dinn*. Differences are boxed in grey. Composed by Lucas Pope. Reduction and analysis by the author. Courtesy Lucas Pope. 130

Figure 4.4: Mapping chord relationships using Riemannian space in "Soldiers of the Sea." Figure by the author. 132

Figure 4.5: Bars 1–20, "Ruins," *Undertale*. Composed by Toby Fox. Reduction and analysis by the author. Courtesy Toby Fox. 135

Figure 4.6: Formal sections in "Ruins." Figure by the author. 137

Figure 4.7: Motivic repetition in bars 1–16, "Ghost Fight," *Undertale*. Composed by Toby Fox. Reduction and analysis by the author. Courtesy Toby Fox. 138

Figure 4.8: Screenshot of the Napstablook battle scene, *Undertale*. Courtesy Toby Fox. 139

Figure 4.9: Screenshot of the protagonist visiting Napstablook's home, *Undertale*. Courtesy Toby Fox. 140

Figure 4.10: "Your Best Friend," *Undertale*. Composed by Toby Fox. Transcription by the author. Courtesy Toby Fox. 143

Figure 4.11: Soundscape analysis of Flowey tutorial scene, *Undertale*. Figure by the author. 144

Figure 4.12: Analysis, Dénouement scene, *Undertale*. Figure by the author. 145

Figure 4.13: Screenshots showing three visual styles during the Flowey boss fight, *Undertale*. Courtesy Toby Fox. 146

Figure 5.1: One possible model of authenticity. Figure by the author. 154

Figure 5.2: Bars 1–17, "Carousel," *Detroit: Become Human*. Composed by Philip Sheppard. Reduction and analysis by the author. Original music copyright Sony Interactive Entertainment and Quantic Dream. 158

Figure 5.3: Bars 1–8, "Sloprano," *Conker's Bad Fur Day*. Composed by Robin Beanland. Reduction and analysis by the author. Original music copyright Rare. 160

Figure 5.4: Bars 1–8, "Votre toast, je peux vous le rendre"
(Toreador Song), *Carmen*. Composed by Georges Bizet.
Public domain. Analysis by the author. 160

Figure 5.5: Screenshot, encounter with the Great Mighty Poo,
Conker's Bad Fur Day. Reproduced by permission of Rare. 161

Figure 5.6: Repetition in "Glory to Arstotzka," *Papers, Please*.
Composed by Lucas Pope. Reduction and analysis by the author.
Courtesy Lucas Pope. 164

Figure 5.7: Motivic development in the "Death Theme," *Papers,
Please*. Composed by Lucas Pope. Reduction and analysis by
the author. Courtesy Lucas Pope. 165

Figure 5.8: Bars 1–12, "Some Place We Called Home," *This War
of Mine*. Note the added tones in the basic chord structure and the
changes in meter. Composed by Piotr Musiał and Krzysztof Lipka.
Reduction and analysis by the author. Courtesy 11 bit studios. 166

Acknowledgments

This research was funded by a grant from the Social Sciences and Humanities Research Council of Canada, and by the Faculty of Arts and Science, Queen's University.

I write this today, nearing the end of a global pandemic, with a new perspective on digital content and our society's ever-changing relationship to it in a context that I hoped I would never have to face. We have all had a very difficult two years, and I can't begin to understand what all of you reading this book have gone through, but writing a book on digital media at the same time as *using* digital media for basically every aspect of my job for over a year is a shared experience that I think has transformed us all. One of the most surprising things I faced at the beginning of the pandemic, though, was logging in to my Music and Video Games course for the first time over Zoom and realizing that I was in a (virtual) room with one hundred young people who, through their previous gaming experience, were well versed in tools such as Discord and Steam that allowed them to transition to online learning more easily than their peers. For that, I was infinitely grateful.

This one-year project turned into a two-year one due to having to develop online course content essentially from scratch, and I have to begin by extending an extremely warm thank-you to Courtney Morales, my editor, and all the other staff at Lexington Press. Courtney in particular has shown infinite patience during the numerous delays brought on by the pandemic and my own inexperience writing a book for the first time, and has been more than generous in answering my fine-detail questions during an ever-changing time. I cannot thank you enough.

I reached out to several of the game authors, composers, and publishers to secure copyright permission for the works included in this volume and received a warm reception from practically everyone I contacted. Special thanks go out (in no particular order as everyone was equally wonderful!) to Toby Fox, Lucas Pope, Konrad and the rest of the team at 11 bit studios; Dimitris Ilias, Giannis Georgantelis, and Simon Fournier at Ubisoft; and finally the marketing team at Electronic Arts. I appreciate all of your feedback and your enthusiasm for the project, as well as your willingness to work with me as I revised scores.

I am always humbled and amazed by the collegiality and generosity of my professional colleagues in the sphere of ludomusicology, and particularly those that I have met over the last five years at the North American Conference on Video Game Music. From comments on presentations to discussions over coffee and generous follow-up after the conference on Discord and via e-mail, their willingness to collaborate and give feedback has been a gift in writing this book. These wonderful scholars are crafting a new field of study as we speak, one that goes beyond the page, which is (to borrow a phrase I use later in this book) infinitely exciting to me! However, I would like to acknowledge a few people by name. Gregg Rossetti and Dan Donnelly gave invaluable feedback on the Greek music chapter that helped me to refine several ideas and put the scholarship into better context. Tim Summers provided some lovely feedback during coffee breaks and a few presentations that inspired a few ideas on adapting the epic. But above all, I would like to thank Karen Cook, who provided me with a wealth of resources on medievalisms, historicism in game sound, tropes, and a variety of other topics throughout this book. Karen, this book is 200 percent better than what I started with based on your feedback and you have provided me with an incredibly solid foundation in your own research on which to build this work.

Here at Queen's University I have also had many lovely colleagues who have encouraged and supported me in my research and teaching work. I would therefore like to extend a thank-you to Margaret Walker, who mentored me in my transition from adjunct faculty to tenure-track and through my first book-writing experience with suggestions about proposal formats, chapter structure, and a wealth of other knowledge. Colleen Renihan has inspired me daily with her positivity, astuteness to the social responsibility of music, and her general support for my work, allowing me to bounce ideas off her and supporting me at conferences. Darrell Christie has supported me with secret whiteboard messages of awesomeness on days where I needed it the most, but she has also reached out for collaborations in our teaching and performance activities, actively embracing video game music as a new repertoire for his choral students. John Burge has been my biggest advocate since he hosted me upon my arrival to Kingston in 2008 and has promoted my professional development since that time in a myriad of ways that demonstrate his generosity, love of music, and engagement with the community on multiple levels. Jenn Stephenson, who I met upon the merger of our two departments six years ago, has proved to be my Drama Doppelgänger in a way that has allowed us to bounce ideas about play, game, immersion, and magic circle off one another and has motivated me to rise to a new level in this work. And Adrian Kelly, research project advisor in our Faculty of Arts and Science, was the impetus for staring this project in the first place, offering valuable advice about writing and applying for funding. I am sure I am missing many other names here, so I'll also just briefly say thanks to Cindy, Julia, Katherine, Kelsey, Kim, Kornel, and Peter for all your patience in dealing with me over the last few years! You are the loveliest of people and colleagues. Outside of Queen's, I am also grateful for the wonderful experience I had in grad school and how it prepared me to move forth from my PhD research into a very different area of study. My supervisor, John Roeder, was instrumental in this, and I have continued to refer back to his own research as I have moved into gaming (as you'll see in the introduction). John taught me to work systematically, question my own analyses and assumptions, and think globally.

Lastly, I would like to wrap up with some personal acknowledgements. To Sheila, my longest friend: thank you for your positivity, compassion to the world, and willingness to randomly deliver gifts of baked goods. To Chandra, thanks for your enthusiasm at trying out new cooperative party games, your fresh perspective on media forms, and just being you in general. But most importantly, these acknowledgements wouldn't be complete without the biggest thank-you to Adrian, my long-term partner who became my husband in the middle of this project. From tolerating my weird work hours to setting up my home office to taking care of dinner when I just couldn't balance one more thing on top of online teaching, I couldn't have done any of this without you. To paraphrase your grandmother, the love we share and the teamwork we have built out of this have made us both stronger. Now, if I could only get you to play *Dark Souls* without trying to murder the most important NPC in the first five minutes of the game, maybe then I'll finally get to watch a playthrough and truly judge if it's the authentic experience everyone says it is. I can dream, right?

Introduction

UNDERSTANDING GAMES AS A CULTURAL CONSTRUCTION

In a recent conversation with a colleague, I expressed frustration with the way in which music theory as a discipline is often depicted: stodgy, rule-following, and an artifact of the past infinitely destined to be looking backwards rather than forward. The discipline certainly deserves that reputation in some ways, with recent discussions by Philip Ewell and others on music theory's white racial frame highlighting a strong streak of musical (and social) conservatism among its objectors.[1] However, in engaging with the ludomusicology community I have experienced the opposite of that conservatism in many ways. These scholars are interested in breaking new ground on how we understand and perceive music based not on the rules of the past, but rather in the here and now. How does video game music impact player decisions? How does it reflect (or refute) and create (or limit) their identity as both gamers and more broadly as human beings? How does their understanding of video games and their music impact their understanding of day-to-day lived cultural experiences? We are in the midst of creating a new, ever-changing digital popular culture shaped simultaneously by both its creators and its audiences.

This approach to music theory grounded in the here and now is infinitely exciting to me. One of the most fascinating aspects of being a music theorist is, in my view, not in creating a series of rules but rather is understanding musics in their particular contexts: how might a musician adjust their performance to highlight a specific impactful moment of the music? How would a folk musician understand (and perform) a specific chord progression differently from a Western-trained classical musician? What impact would certain creative choices make on an audience? Far from being binary choices, these questions speak to the fact that we as creators and listeners understand music as a unique, individualized experience based on an infinite variety of different factors from our own personal histories. And in many ways, playing video games is a similar experience.

My upbringing as a female, Canadian music theorist who grew up in the 1980s, for example, might radically change how I understand the plot and interpersonal dynamics between characters in the game, how I listen for and to music and sound, and whether I perceive "Easter eggs" or other more subtle references to earlier source materials. On the other hand, I have also at times noticed a disconnect with the games I play because of such features. My lack of coordination, for example, necessitates a slow, sniper-based approach in first-person shooters that is dramatically at odds with the typically fast-paced, loud, and rhythmically propulsive music that accompanies many games of that genre. This individualized mode of perception has been particularly apparent to me since I started teaching an undergraduate course on music and video games. While I grew up on the cusp of the digital and analog ages, seeing the development of early game consoles throughout my childhood but still familiar with now-obsolete forms of technology such as record players, film reels, and five-digit telephone numbers, my students have grown up solely in the midst of a purely digital culture.

One example that particularly struck me recently was a class discussion of *Cuphead*, released in 2017. For those of you not familiar with the game, *Cuphead* is strikingly modelled on the animation style of the 1920s through 1940s, evoking many of the characters and scenes of animation shorts from that era, and also imitating that time period through choices in the vocabulary of its dialogue and superimposed sound and visual effects that mimic reel-to-reel film playback. My students cited many features of its music and gameplay that I also found effective, including its pacing, simple modes of interaction, and unique sense of style, but there were many elements that I responded to completely differently than my students. For example, while many of the historical cartoons referenced in the game still aired as reruns on television during my childhood in the 1980s, their problematic racial and gender stereotyping mean that they are no longer acceptable to air today.[2] In my gameplay, nostalgia was evoked through my specific memories of the source materials, their music, the game's uncanny accuracy to specific cartoons (which I was well familiar with through repeated Saturday-morning viewings as a child), and its overlayering of a static effect that reminded me of playing records and of old classroom film reels. For my Gen Z students, on the other hand, the game functioned not as a nostalgic memory cue to the 1980s, but instead as a retro-genre game because of its ample use of big band jazz, muted colour palette, and animation resemblances to Mickey Mouse (the character they were most familiar with from the era). We perceived the game as two different creations: mine was a historical reenactment, while theirs was a new creative work in a niche genre. And yet somehow both groups came out of the conversation agreeing that the game felt *authentic*, by whatever measure we were using.

However, despite these differences in perception, it was indisputable that we had a shared culture of video games. In spite of having different degrees (and eras) of experience with video games, we all had a shared set of expectations about how to engage with them, ranging from our physical interfaces (standard key and button mappings for jump and fire, for example) to narrative tropes (such as rescuing the princess from the castle) to the behaviour of particular character archetypes (your troubled companion will eventually turn to the dark side, so watch your back!). No less was true of sound and music in games. Consonant sounds cued us to moments of success without stating it explicitly, dissonant sounds and volume swells anticipated horrors to come, and even the lack of music functioned as a cue to the player, signalling that "nothing's happening here" in an open-world adventure game or conversely creating a cue to listen intently for imminent enemies in a horror game.

Some of our most enjoyable gaming experiences have come from the intentional undermining of these musical expectations. Karen Cook, for example, examines the subversion of the "rescuing the princess" trope in the satirical *The Bard's Tale* (2004), outlining that not only the tropes but also other elements of narrative play "subtly with preconceived expectations about RPGS, and other less overt sonic and musical elements act toward the same end."[3] Our shared cultural references through online media cultures in the twenty-first century were a rich source of information about games and narrative, but also a quick lexicon for game designers to reference and subvert expectations, as I discovered recently while encountering "The cake is a lie" scrawled on a wall during my playthrough of indie developer Kiary Games' *Tiny Room Stories: Town Mystery* (or was that mega-publisher Valve Studios' *Portal*?). More and more in the twenty-first century, audiences are very aware of (and directly participate in) crossovers between different media. Not only are audiences consumers, intaking games, movies, television, and music created by others, but they are also creators, bringing new life to these works by producing memes, social media videos, online discussions, and more. The border between media forms is a porous one at best, but nevertheless a common digital culture is taking shape. As a result, we have set expectations for how our games are going to look, feel, and sound.

This book is about these gaming expectations, but more specifically how players perceive and respond to the expectations created by music and other sounds. This has been a frequent subject of discussion in ludomusicology, but this book will examine the question with a key difference. As a music theorist, I'm ultimately interested in how our perception and understanding of *musical/sonic structure* impacts our gaming experience. I am also interested in some of the unique elements of gameplay in contrast to other forms of media: the reference to players' out-of-game knowledge, the construction of fictional worlds, the interaction of music with other components of the overall soundscape, and the impact of interactivity. As many scholars have argued, video games as a medium differ significantly from other media that integrate visuals and sound. Interactivity—the ability of the player to actively make choices that impact the outcome of gameplay—is cited as one major difference, but the implication of this is that player choice changes gameplay and therefore no two players experience a game identically.[4] From an analytical perspective, this has meant a radical rethink in how to approach the musical work.

Isabella van Elferen discusses several of these aspects in her ALI (affect literacy interaction) model of immersion, indicating that player involvement blends three components: the emotionally oriented affect, memory-based media *literacy* (what I will generally call *intertextuality* in this book), and player/game cognitive and physical *interaction*.[5] Her focus is on how this impacts players' perception of sound in games, and she argues that much of the way that sound and music communicate these three components in-game occurs unconsciously, often at an emotional level. Furthermore, while musical meaning is highly subjective, in video games designers and composers often want to evoke more predictable responses despite differences in players' backgrounds.[6] There is a lot to unpack here. Music/sound in games engages with both emotion and memory but may also act as a cue to action or as another layer to the narrative exposition. Van Elferen is not the only scholar to examine the impact of intertextuality. Tim Summers, for example, presents intertextuality as one of many possible analytical foci in game music, grouping it with semiotics and topic analysis and stating that "game music does not exist in a musically sealed world, but draws upon a common musical lexicon from broader culture. Analysing the musical-rhetorical gestures and musical signs in games helps to understand the ways in which musical meanings are configured and re-configured in games."[7] William Gibbons discusses the specific use of preexisting classical music as intertextual reference in *Unlimited Replays*, and interestingly observes that the impact of preexisting reference material isn't just a one-way relationship: "Instead of depending on players' prior knowledge, the music and gameplay work together to create a kind of feedback loop: the games teach players to associate the sound of the music with the era of the game, while the overtly historical (that is, 'old') sound of the music encourages a sense of immersion."[8] Furthermore, Gibbons observes that these associations don't need to be historically accurate. Since players don't generally have a detailed, accurate knowledge of the historical past, any music that sounds old to the player might be used to evoke the past.[9]

AUTHENTICITY

This book will explore these various facets of player experience with music. However, at the core of the discussion I will keep returning to one term: authenticity. What is authenticity? Literally, we might expect it to mean conforming to an original form, to reality, or to fact. But this is not the way the term is understood among gamers. Authenticity is much more about how gameplay matches the player's expectations. Jesper Juul, in *Handmade Pixels: Independent Video Games and the Quest for Authenticity*, explores this subject in depth, particularly relating it to the rise of indie game development in the twenty-first century (a subject that will be explored in chapter 4). He emphasizes a link to the indie gaming ethos of independence, small budgets, and hand-crafted projects by a small team of designers and resultingly asks, "But what is an authentic game? Authenticity usually refers to the absence of a range of ills: selling out, being unoriginal, being controlled by money, being superficial, or angling for fame."[10] This definition is strongly impacted by Juul's focus on the indie mindset and centers on the perception of a game as a creative work. However, players use the term *authenticity* in a much wider context not necessarily focused on this game-as-artistic-product perspective, sometimes referring to historical accuracy, while at other times referring to a nebulous, hard-to-describe feeling of whether the various components of the game appear to form a coherent, integrated whole. Juul includes one perspective on authenticity from Richard Peterson, a scholar on country music, that I find particularly apt given this multiplicity of definitions: "Authenticity is not inherent in the object or event that is designated authentic but is a socially agreed upon construct in which the past is to a degree misremembered."[11] A game need not be accurate in order to be perceived as authentic by its players. As Elizabeth Randall Upton describes in regard to Early Music performance, the sense of authenticity is reliant not only on historical accuracy but also "on a complex interaction of experiences and expectations on the part of the audience."[12]

Richard Stevens and Dave Raybould aptly speak to this definition of authentic as including the real (such as whether the game represents historically accurate events), but their definition also includes what they term the "mediated real," how games "match a player's expectations or schema."[13] In their analysis of *Battlefield 4*, they identify (like Tim Summers) that players' perception of sounds provides a means of creating materiality to the game experience.[14] This materiality, in my view, can occur through modifications to sound, such as reverberation that might support players' proprioception (the perception of one's position and movement within the virtual space) or other forms of perceiving physicality in the gameworld (a literal definition of the *real*), but also through music that supports the emotional landscape of the narrative and through player reactions.

One way to approach this disparity might be to separate these various ways of understanding authenticity explicitly. Felix Zimmermann takes one such approach, breaking the concept down into *object authenticity* (in our case, this might be understood as the accuracy of the game and its narrative, as well as any historical music) and *subject authenticity* (in our case, the manner in which players react to their game experiences).[15] Michał Mochocki takes a similar approach, acknowledging that *accuracy* and *authenticity* are often conflated and suggesting that perhaps separating the concepts of accuracy to facts versus accuracy to feelings (what I believe gamers are often referring to when using the term authenticity) might better solve this confusion. He astutely goes one step further, pulling in scholarship by Ning Wang that breaks authenticity into three components: 1) *objective authenticity* (historically verifiable information), 2) *constructive authenticity* (player expectations and projections), and 3) *existential authenticity* (player feelings).[16] The case studies I will present in this book will touch on all three of these modes of authenticity.

These distinctions help to mitigate authenticity's sticky relationship with historical accuracy, a concept that will be examined throughout this book in terms of the ancient historical past, player media literacy based on previous experiences, musical memory, and the rise of retro gaming culture. In each case, I will discuss the impact of this tension between the real and the remembered, the literal past and the constructed past. Felix Zimmermann aptly observes that:

> I am intentionally not talking about history but about the Past. I want to stress that history, as a scientific but also pop-cultural or societal narration about past time-spaces, has a complicated relationship to authenticity. People turn to the Past to find what they think they are lacking in their everyday lives. But does that mean they also turn to history? Or are they creating their own histories in their search for authenticity? Of what kind are these histories?[17]

I would argue that players absolutely create their own histories. Sometimes this occurs literally. When playing the *Civilization* series, for example, the outcome of the game is not determined by the results of historical battles of the past but instead by the player's skill in navigating resource management and conflict. Its music, as Karen Cook identifies, signifies and reinforces the player's progress and "is thus uniquely tailored to the pace of each individual game; as a result, the soundtracks reinforce the player's immersion into the game by creating a player-centric sound world and alert the player to the passage of time, both in and out of the game."[18] But this creation of history may also occur more figuratively, with players' own memories acting as filters or sources of mediation.

This tension is explored in chapter 1 through the lens of gameworld, where I will begin by examining objective authenticity in a case study of two games invoking Ancient Greece: *Assassin's Creed: Odyssey* and *Hades*. I will use score transcriptions to more closely examine Greek-inspired features of the music such as additive rhythms and modes, but I will also discuss how these elements have more to do with modern-day players' perception of the musically ancient rather than historical accuracy. This will serve as a launching-off point for discussing how historically based games evoke the past in ways both real and imagined, bringing up questions of whether fantasy can be more authentic than reality (and where we might draw the line between the two). The second half of the chapter will discuss how sound and music can evoke constructive and existential authenticity, forms of authenticity that speak to player perception and emotion, through their depiction of narrative and game cues. The musical analysis in this portion of the chapter will shift to a soundscape-inspired timeline analysis to highlight interactions between sound, music, visuals, game functions, and narrative that serve to create more immersive and effective gameplay.

Chapter 2 explores a similar tension but with a stronger focus on constructive authenticity, examining how players build their perception and expectation of gameplay through their knowledge of previous media. Centered around the use of tropes in narrative, the chapter begins by examining previous theories by film music scholars such as David Neumeyer and Robert Hatten, and then expands the definition of trope to better encompass modern interactive media forms and online communities. Tropes, both musically and narratively, help to structure player expectations of gameplay and to create a parallel emotional universe, but are only effective if the player is familiar with the source references they invoke. Case study analyses will focus on musical medievalisms in Fantasy games,[19] a common source of topic and trope cited by ludomusicologists such as Karen Cook and James Cook, and will discuss how features of historical medieval and baroque music are adapted for this means of representation. However, the tension between objective and constructive authenticity comes to the forefront given the discrepancies between fantasy and reality, and the implications that this creates for players' reconceptualization of history. The final portion of the chapter will focus on tropes that reference more modern source materials and the problems that arise from this form of cultural mediation in games such as *Samba de Amigo*.

Chapter 3 takes a constructive authenticity perspective, examining how musical motives communicate and whether that provides a barrier to immersion and authenticity. Leitmotif is one of the best-known means of musical representation, but as Matthew Bribitzer-Stull and other scholars examine, other forms of *associative themes* allow for the representation of character, dramatic events, and other elements of the fictional universe. The discussion will break down the musical examples into three categories—thematic association through character, through gameworld (linking to previous discussions in chapter 1), and through developmental Leitmotif. Scholars argue that a simple re-presentation of musical theme lacks an aspect of musical development required for a true definition of Leitmotif; however, many games use literal (or close to literal) repetition of musical themes to help create coherence between sequels within the same series. This suggests two different functions for thematic association in gaming: one simply as a simple semiotic marker, and the other more closely linked to depicting developments in narrative. Examples from the *Street Fighter*, *The Legend of Zelda*, *Star Wars*, *Halo*, and *Sonic the Hedgehog* series will help to clarify these two different processes, and strategies of motivic analysis will show how musical variation becomes an important component in motivic evolution.

Lastly, chapter 4 takes the perspective of constructive and existential authenticity, examining how player expectation and player emotions feed upon one another as well as upon memory and nostalgia. Retro games have grown in popularity since the turn of the millennium. These games are not created through strict replications of past technologies, but instead use modern tools to create new works that evoke a general sense of the past. The aim is to match the overall expectations of gaming in the past (e.g., 16-bit sound) while filtering out some of the uglier, less enjoyable elements (such as long loading times) that players may have forgotten over time. The chapter will begin by discussing the growing appeal of retro gaming and its origins in the indie gaming world, an origin that produces yet another definition for authenticity based on the game as an artistic, independent, creative work. From a musical analysis standpoint, my definition of retro and its musical features will pull from work by Karen Collins, Nikita Braguinski, and Andra Ivănescu, serving as a launching-off point for more in-depth analyses of music and sound from *Return of the Obra Dinn* and *Undertale*. The analyses, which focus not only on pitch and rhythm but also on timbre, narrative events, and visual components, will highlight how modern retro blends the sounds of past and present technologies, an interaction that allows for a critique and adaptation of the past that combines all three modes of authenticity. The shift between congruence and incongruence of gaming expectations will be a particular focus of this chapter and will set up our final discussion in the conclusion: Are these three modes of authenticity compatible in a single definition of the term? How does the game effectively communicate to the player, and how might the player be pushed towards making particular choices through sound?

The Role of Virtuality

An important concept in discussing players' understandings of authenticity is the *magic circle*. According to van Elferen, the magic circle is "the accumulation of game rules, gameplay, and immersion that delineate the game experience as a separate space outside the ordinary world," and she argues that music has the ability to expand this magic circle of gaming into the real-life surroundings of the player.[20] She also suggests that music is perceived through a level of abstraction that might be understood in the same way as hyperreality and consequently manifests many of the same properties as our mediated understanding of authenticity in games.

> It is a form of virtual reality created by a medium—music—but one that is much less easily distinguished from real life than that created by other media; this is because of its rootedness in *collective and personal emotions and connotations,* on the one hand, and the necessary temporal simultaneity of such musical virtuality with *day-to-day reality,* on the other. In a sense, musical virtuality is neither real nor virtual, whereas the game's half-real is both. It is this musically evoked altered reality that demonstrates how music can reveal the slippery nature of concepts like the magic circle, real life, virtual reality, and hyperreality—concepts that are buzzing through current academic debates, to which musicology can and should contribute valuable insights.[21] *(emphasis my own)*

I have highlighted "collective and personal emotions and connotations" in this quote as I believe it to be identical to our concepts of constructive and existential authenticity. Player experience determines that they understand connotations, and effective renderings of these connotations produce emotional reactions. The half-real (or perhaps more appropriately trans-real, as it continuously crosses the border of the magic circle) status described above is strongly rooted in our perception of sound as a sensory input. While physical actions in gaming involve the mediation of a controller or another form of input, and visuals (with the exception of VR) involve the frame of the screen that restricts our perspective, we perceive sound as our true physical selves, immersing us in the virtual space. It is virtual in the sense that it is digitally created and exists as part of the gameplay experience, but it exists in real life, and thus the duality normally presented between these two oppositions does not work here. Although virtual is often positioned as the opposite of real life, and parallels are sometimes made to digital versus analog, Tim Summers is aware of this ideological conflict and argues that virtual is not the opposite of real but rather exists as a mode of understanding reality. Inspired by Marcel Proust, he suggests that we perceive the digital as real but not concrete, a form of unsensed reality.[22] However, even this compromise requires some tweaking as we *hear* sound and music as our true selves (and thus it is sensed). Regardless of these different perspectives, these scholars agree that music is a common ground between reality and the virtual, bridging the gap and allowing access to our emotions, imagination, and more.

THE ANALYTICAL APPROACH

While scholarly approaches to video game music have previously focused on questions of player reception and perception, only recently has there been a stronger emphasis on examining musical structure, understandable given the variability of game playback and its synchronization of sound. However, several recent scholars have shifted to more of a ludo-music-theory perspective, an approach that I will also be taking in this text. These scholars are aware of the difficulties of using a fixed musical score for a medium that is inherently shifting. Tim Summers, for example, wonders "we may be at something of a loss when, in the absence of a score or a fixed sonic output, our familiar tools of the trade are not available to us. The analyst is thus left to ask, 'What is the object of analysis to be and how is it to be analyzed?'"[23] There is no one answer to this question. While some analysts have integrated score transcriptions as a key component of their practice, others have begun designing new forms of analysis to account for these non-fixed playthroughs. Elizabeth Medina-Gray, for example, has developed a theory of musical modularity that mirrors game music design principles: Music may be activated by particular in-game triggers, and variable looping, layering, and other recombinant processes accommodate gameplay of unspecified duration.[24] Her modular approach allows us to understand these looping processes in conjunction with game events, highlighting one of the most essential components to video game as an art form: It not only involves sound and music, but the narrative and visuals are elements that cannot be separated from the players' game experience. Medina-Gray uses this modular approach to introduce a discussion of smoothness versus disjunction in game sound, both between one modular unit to the next (that is, how connected do consecutive musical units sound compared to one another?) and also between modules that sound simultaneously (how well do musical layers fit with one another in terms of rhythm, tonality, instrumentation, and other elements?). I am adapting several of her ideas in this text. My analysis, like Medina-Gray's, will examine pitch and rhythm, but it will go further to also discuss timbre, instrumentation, and other elements of cohesiveness (or lack thereof). Additionally, smoothness versus disjunction will be a major focus in my work, albeit repositioned as *congruence* versus *incongruence*, terms originally taken from Karen Collins but adapted to accommodate elements beyond the score, including player expectation.[25]

Sean Atkinson's work has inspired my analysis of tropes in chapter 2. In "Soaring Through the Sky: Topics and Tropes in Video Game Music," Atkinson provides a ludomusicological perspective on theories of topic and trope originally established in Western art music and in film music, but to me one of the most insightful parts of his essay is his concrete focus on specific musical elements of the soaring/flight trope. He identifies rising melodic lines, open fifths, prominent leaps, and the Lydian mode as distinct musically observable features of this trope, allowing the reader to pin down exact music structure features that impact the player's perception and imagination of flight.[26] Atkinson also makes the observation that tropes are not solitary signifiers, but work in combination to provide a more nuanced meaning:

> The soaring topic, then, is itself a trope of many other topics: the machine, with its repetitive motions; the martial, with an emphasis on large ascending leaps; transcendence, with emphasis on long sweeping gestures often performed by a harp; and the supernatural, indicated by the presence of the Lydian mode. But given the consistency of its use over the last thirty years in cinema and video games, this onetime trope has seemingly become a topic, one that is instantly recognizable by movie goers and video game players alike as a musical marker of flight through the open air, often by magical or supernatural means.[27]

While he specifically identifies each feature as a topic in this description, I would argue these form combinations of tropes as well. The machine, martial, and supernatural are not inherent musical properties, but rather are references established in previous media genres. And in contrast with the fixed medium of film, Atkinson observes that this use of trope and topic in game "could influence a player's decisions, having a real and meaningful impact on the experience of the game."[28] Thus music structure elements of melody, harmony, consonance/dissonance, and tonality have a direct impact on player action. William Gibbons similarly identifies concrete trope features throughout his discussion of historical recontextualization in *Unlimited Replays*. While I will save most of this discussion for the upcoming chapters on gameworld and trope, I will mention here that Gibbons distinguishes between different levels of recognizability. Some players might recognize harpsichord, dance rhythms, and small orchestra as historical features of baroque music, but some will only understand these as generally evoking the past. Nevertheless, for both groups of players these features (as Gibbons describes) help to "quickly situate the game—and themselves—in a specific place and time."[29] Thus music structure features are important in establishing gameworld, even if the player is only perceiving these at a subconscious rather than conscious level.

Steven Reale also takes a strong music theory approach in his 2016 video analysis of music in *Portal 2*.[30] One of the best features of his analysis, in my view, is the strong connection he makes between compositional choice and the game's narrative. In describing, for example, a process of chord arpeggiation in "9999999," he clearly identifies that "the impression created by the theme is that an algorithmic compositional process has, like GlaDOS herself [the main antagonist], run amok."[31] Reale's overall argument is that the music's developmental processes mirror not only the puzzle structure of the game itself (particularly its iterative aspects) but also the duality of the main antagonist, an artificial intelligence struggling between her original human personality and her amoral artificial mind. Reale clearly links harmonic progression and narrative processes; for instance, the first half of his analysis outlines numerous instances of harmonic duality as well as the opposition of consonance and dissonance, reflecting GlaDOS's duality of natural and artificial. The music thus helps the game tell the story. The second half of his analysis focuses on theme and variation procedures, which once again mirror the gameplay. The player encounters puzzles of ever-increasing difficulty as they move from one level to the next, progressively introducing new mechanics that morph the gameplay parameters and expectations. Reale links level design and musical design, viewing the music of the entire game as a continual process of developing variation. I believe developing variation occurs in more game music than we are aware of. My analyses in chapter 3 include an example of developing variation in *Halo*, where progressive change to the main motive reflects the changing levels of certainty in the protagonist's search for his beloved companion.

Most importantly, though, Reale describes that there are "rich interactions that are possible when music accompanies storytelling."[32] This is a major focus of this book. In video gaming, where we are not listening to music in isolation but rather are constantly experiencing it alongside both visuals and narrative, it is disingenuous to separate music from these other components. Although there are some cases where music acts simply to fill silence, in most games, music is intended to inform the narrative in some sort of way. Thus, some important questions arise: How does music adapt to gameplay of unspecified duration? How does music align with visuals? How do narrative cues impact our perception of music? And how do music and sound engage with one another as two components of a complete soundscape? One of the underlying principles of this tome is that a multimedia approach is necessary for any examination of musical structure in video gaming. This approach will require a variety of analytical strategies, including those adapted from preexisting work in music theory, film music, and ludomusicology, but also new approaches.

Implications of the Analytical Approach

To achieve this multimedia perspective, the analyses in this book will mix traditional, score-based analyses and soundscape analyses that depict sound's interactive role within the game as a whole. While players might listen to soundtracks as standalone musical works after and outside of gameplay, that is not their default mode for encountering this music. Players engage with this music in-game, in its rich context as an audio, visual, and kinetic medium. Currently, few analytical methodologies exist for understanding game music in that larger context. Scholars are aware of the problem, as well as other unique challenges that game music pose to traditional forms of musical analysis. As Steven Reale observes,

> most analytical methodologies rely on treating some kind of fixed musical notation as an input and begin to break down when such notation can be neither acquired nor self-produced through transcription. But since there continues to exist vast amounts of music that can be fixed into notation, there has not generally been felt a widespread, urgent need to rework or revise existing methodologies to accommodate those that cannot.[33]

While Reale's work, as well as that of Elizabeth Medina-Gray, has focused on accommodating the shifting temporality of gaming music, with its variable length and modular construction, my work here will focus on another component of this problem: sound and music as one element of an artistic form that comprises many (nonmusical) dimensions. As van Elferen explains, analyzing music's role in video games as a multimedia form requires a combination of approaches in narrative, ludology, and interactivity.[34] I also believe that sound is just as important as music in this study. Both have an active cueing function in-game and prompt a multitude of player interactions, and in many cases they are indistinguishable from one another within a continuum between sound and music.

However, as this is a book focusing on musical structure, I did want to speak to the role of musical analysis itself. Players, when engaging with games, understand music as part of the whole. While they might explicitly analyze the music in games that require a certain amount of predictive ability in relation to musical input (such as rhythm games), for the most part players do not consciously analyze a video game's music when they play. However, I would argue that despite this lack of conscious activity, players do *subconsciously* detect patterns within sound and music—most players, for example, would easily identify a change in music that occurs when moving from one in-game area to another—and that perceiving these patterns and their correspondences with other elements of the game augments their feelings of authenticity regarding the game. This, consequently, contributes to player satisfaction, immersion, and flow.

Players thus subconsciously listen for patterns, but this does not require explicit understanding or training in music theory. Patterns in sound tell the player a wide variety of information about the game. Repetition, for instance, indicates a maintenance of the status-quo and might thereby inform the player of a safety or danger state. The repetition of a single sound might be paired with a specific action on the part of the player, such as the well-known *cha-ching* sound that occurs when Mario picks up a coin, or the sound of gunfire heard when the player in a platformer or shooter fires their weapon. Such a sound is *cueing* the player that their gesture has been realized successfully. Sometimes a similar function occurs with implications for the future game state. Players can predict upcoming events through changes to the status quo in the music. Iain Hart, for example, cites *Skyrim*'s use of a musical cue to indicate detection by an enemy even when that enemy is not visible to the player, and states that players learn to use this predictive strategically to plan stealth attacks.[35] In Hart's example, the change to the present musical state is *evoking* new actions or events to come, and the predictability of hearing the same enemy song multiple times in the game allows the player to understand the musical theme's semantic meaning and to plan future actions accordingly.

Aural patterns also help to frame the game in terms of virtual geographies (for example, the move from one gameworld location to another), time (such as *Super Mario*'s doubling of the musical tempo when the player's timer is running out), and modes of gameplay (including shifts between cut scene, interactive dialogue, and active gameplay), a function that I will term *framing*. Players also refer to both current sounds and previous knowledge to understand how these patterns fill in detail to create a rich, detailed gameworld. Tim Summers identifies this as *texturing*, which helps to enhance the game experience.[36]

Figure 0.1 shows these four functions of game sound used by players to interpret patterns from sound as sensory input. Texturing and framing both have strong connections to gameworld and the aspects that we understand to exist inside the magic circle, elaborating the narrative and backstory and establishing the world as a complex, varied space that can be virtually navigated. These manifest what Michel Chion might term *semantic listening*—that is, sound that communicates meaning through an associated code. Cueing and evoking relate to the ways that the player engages with and creates actions in the game (what Chion might term *causal listening*), and thus focus on the crossing of the magic circle boundary.[37]

The Role of Transcriptions

A common approach to musical analyses of video game music is to work with published scores or analyst-produced transcriptions. However, such transcriptions or scores are problematic in several ways. Representing

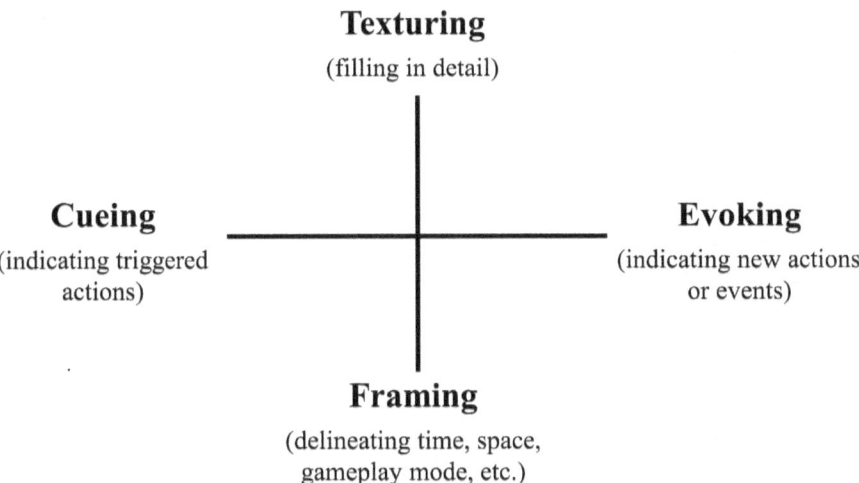

Figure 0.1. Four functions for game sound. *Source*: **Figure by the author.**

gameplay of indefinite duration, for example, with a single fixed-duration score loses some detail in how looping and other procedures accommodate variability in length. Looping music additionally also involves a process of layering not depicted in most score transcriptions. But more importantly, transcriptions that are based on Western musical notation tend to restrict themselves to representations of pitch and rhythm, with minimal detail about other components. I have been mulling over this issue since my graduate school days, where I had the good fortune to take courses on transcription and its role in ethnomusicology with John Roeder and Michael Tenzer. A major component of our seminar discussion was the subjectivity of transcription. Transcriptions focus on the elements that the transcriber deems most important, which may differ significantly from the perception of the work by performers and other practitioners. As John Roeder explains in his introduction to *Analytical and Cross-Cultural Studies in World Music*, "in seeking to describe and compare music, one naturally wants to concentrate on essential features, but, depending on the nature of the music, these may be represented with varying definiteness and fidelity."[38] Roeder further goes on to state:

> Comparisons are complicated not only by these different ontologies, but also by the subjectivity of transcription. The information on a score, representing the composer's performance instructions, does not comport readily with a notation of what the analyst hears, even if informed by the creators. On the one hand, unnotated performance practices may greatly affect how a score sounds. . . . On the other hand, transcriptional accuracy may be limited by failing to discriminate significant differences, by imposing foreign conceptions of pitch and time, or by the limitations of the recording itself.[39]

Gaming has a clear parallel: the analyst's transcription may not correspond to the original composer's intention of meter, for example, even if producing a representation that might sound identical upon playback. Unnotated performance practices might include player reactions, accompanying visuals, narrative prompts that impact player decisions (and thus score playback), and a wide range of other factors. Traditional scores do not represent these components, and thus for an analytical approach based on the integration of visual, narrative, and auditory components, standard transcriptions are lacking somewhat in the way that they can illustrate these connections.

That said, as an analyst I will still rely on transcriptions in many of my discussions. However, these transcriptions should not be viewed as representative of all elements of the music in the game and will not be limited merely to rhythm and pitch. In chapter 1, for example, my analysis of Giannis Georgantelis's Greek sea shanties from *Assassin's Creed: Odyssey* is based on only one of two recordings (either the men's or women's choir version) available within the game. Even if focusing only on pitch, rhythm, and lyrics (as I restrict myself to in those examples), there are still discrepancies between the two performances: changes in timing, in the accuracy of pitch, and in the ways that the text is articulated. The songs were intentionally recorded by two different Greek-community choirs, and these discrepancies add to the sense of constructive authenticity for the player as work songs within the game (in other words, we would not expect a group of sailors to sing a work song in pitch-perfect and rhythm-perfect virtuosity, nor would we expect each rendition to sound identical!). Another transcription in chapter 1 focuses on volume changes in *Metro: Exodus* and contains no indication of pitch or rhythm whatsoever since the soundscape at this point in the game is not musical in the traditional sense. The annotation on the figure instead demarcates important shifts from one room to the next within the game, of acousmatic sounds, and of player-generated sounds that might impact game events. In this case the relationship between sound and action is the most important analytical observation. In my analysis of *Undertale's* major boss fight in chapter 4, while there are clear distinctions of pitch and rhythm in the accompanying music, the more important aspect of the scene is the shift between congruence and incongruence with player expectations. As a result, my transcription notates game events and shifts of gameplay mode (between cut scene, free-roam gameplay, and battle) to better understand the alignment of these components with shifts between digital and acoustic timbres. In analyses where I focus on more traditional elements of Western music notation, these are not meant to be the end goal but instead are merely a tool to illuminate elements of harmony, form, phrasing, and their synchronization (or lack thereof) with other multimedia elements. In all cases, the transcriptions should be viewed as representations of my own hearing and my own gameplay, which may differ significantly from yours.

Foundations in Music Theory

Transcriptions are one tool, but what others might we need in order to conceptualize the interaction of musical structure and game forms? Each chapter of this book will pull specific tools into the topic at hand, but the following overarching concepts will serve as a useful beginning for our understanding of musical structure in video games.

1. Tonality Has Implications for Player Perception One of the most basic dualities in describing music in Western culture is that of major versus minor scale. At its simplest level, major and minor are often associated with positive and negative emotions. As David Huron identifies, the major mode has become associated with happiness and "for whatever reason, the minor scale has become associated with sadness. ... At face value, there is nothing inherently 'sad' about the Western minor scale. The most-commonly held view is that the relationship between the minor mode and sadness arises from a learned association."[40] While this simplistic means of emotional association is often avoided by music theorists, who resist such a black-and-white interpretation of basic musical organization, players and game designers nevertheless cite this as a feature affecting their emotional perception of music. Blogger Joe Gilliver, for instance, states that "the reason why we try to pick major and minor keys for compositions is for the emotional feel. Essentially the harmony of the chords and the choices of notes in the melody portrays the feel and emotion of the piece. We associate 'major' as a happy feel and 'minor' as a sad vibe."[41] Given this clear perception on the part of gamers, it is no surprise, as Huron outlines, that "a composer might make use of the minor key merely because its use in a particular context conforms to an existing musical tradition or convention."[42]

This basic opposition of major and minor, however, is not the only means through which scale or mode can impact the player's perception of musical communication. The use of diatonic modes beyond the major and minor scales in video gaming is a pretty significant way of signalling the "other" musically, whether that be an allusion to medieval music, the musically ancient, or locales outside of North American popular culture or, conversely, certain popular music styles. These modes may not be accurate to the cultural musical practices being alluded to, but nevertheless among players these sonorities are understood as evoking something foreign to their default major/minor expectations. As such, the diatonic modes (Ionian, Dorian, Phrygian, Lydian, Mixolydian, Aeolian, and Locrian) and their differences from the major/minor scales will be discussed in several places in this text, including chapter 1's discussion of gameworld and chapter 2's discussion of trope.[43] Note that in this text, despite the fact that Aeolian mode is often considered identical to the natural minor scale, I will show preference for the term *Aeolian* over *natural minor* in contexts where I am referring more broadly to mode and other contexts that avoid the raised leading tone of the minor scale.

Harmonic function will also be a useful perspective to structure our listening expectations. While musical expectation is widely dependent on culture (i.e., the music that listeners are exposed to over the course of their lifetime), in the systems of musical organization most common in North America and Europe a few specific patterns generate particular expectations among their listeners. Tonic chords typically include scale-degree 1 of the key and produce tonal stability. Dominant chords, which typically include the seventh note of the scale, produce a sense of tension in the listener given the proximity of this tone to the tonic note of the scale and its expected resolution there (although this may be weakened depending on the distance from scale-degree 7 to 1—major mode, for instance, features a semitone in this position, while Dorian mode features a whole tone). Subdominant chords, which typically include the sixth and sometimes the fourth notes of the scale, neither reinforce the key nor produce tension; they act as connecting chords. While the classical music of the eighteenth and nineteenth centuries featured the common paradigm of tonic–subdominant–dominant–tonic, modern popular styles and musical styles meant to evoke other cultures often avoid this paradigm. Thus harmonic function and its ordering may serve as a marker of historical (objective) authenticity (such as the examination of baroque-modelled counterpoint in chapter 2), of player expectation of otherness, or as a means of manipulating tension in the player (both constructive and existential authenticity). Huron identifies other components of melody, scale, and harmony that likewise affect player perception of melodic regularities versus irregularities, all of which will be useful for contextualizing our discussion of player expectation: pitch proximity, step declination (with large intervals more likely to ascend and smaller ones to descend), step inertia, melodic regression (the change of direction after a leap), and melodic arch.[44]

2. Cadences Can Occur Independent of Harmony Huron also cites cadence as a musical feature that allows the listener to structure their perception of predictability versus unpredictability, stating that cadences are more predictable than other segments of the music, and that they are often followed by areas of increased musical uncertainty.[45] This suggests that cadence can serve as a useful point of comparison for congruence and incongruence of player expectation in game music. However, I would like to take a much broader approach to cadence than that often given by common-practice harmony. Cadences in this book will not necessarily be determined by harmonic progression, but rather will be understood as a point of musical arrival. Such arrivals might be produced in tonal music through the tension-and-release chord progression V–I (as per the traditional definition of the perfect authentic and imperfect authentic cadences), but they can also be produced in other ways that we often fail to acknowledge. This might include longer note durations, step motion to a foundational note such as the tonic of the key, resolution of a tension tone within a key, or completion of a line of vocal text.

From an atonal perspective, though, these moments of punctuation are created from dramatically different factors. The survival horror genre, for example, frequently uses sudden increases of volume (what some scholars term *nonlinear sound*, although I will avoid that term in this book due to its multiplicity of meanings)[46] within musical scorings that are sparse and dissonant to signal moments of importance such as sudden enemy attacks. The opening cut scene of *Assassin's Creed: Odyssey*, on the other hand, does the opposite, using sudden silence after an increase of texture and volume to signal the moment of arrival where the player must stop watching the cut scene and shift to taking control of the gameplay. The Master Chief arrival in *Halo 3* uses a similar silence-after-arrival effect to its cadence, in this case paired with a shift from dissonance within a thick orchestral scoring to a consonant, thinner-textured piano. These clearly act as moments of musical arrival, regardless of whether they fit the traditional paradigm of V–I harmonic cadences.

3. *Repetition Creates Common Simple Form Structures* Repetition is a key component of video game music due to its early origins in technological limitations, with looping in particular occurring as one major stylistic feature. Loops (repeating segments of music) were used in early video game design as they allowed the programmer to use fewer memory resources for music. The same stored music could be reused over and over again, which from a digital perspective simply re-referenced the memory block where this information was stored. While these programming constraints widened out in the mid-1990s as computer memory capabilities increased, looping as a musical feature was retained. Loops are useful well beyond their memory-saving advantage. In current adaptive audio design, musical loops are frequently used in game areas where the player may be for an undetermined length of time as the music may repeat infinitely until the player decides to move to a new location. This will be a major focus in my analysis of retro gaming styles in chapter 4, but it is useful to also note that looping has, over the last forty years, become an emblematic (and thus expected) element of game music style.

How might we discuss repetitive procedures such as looping? Some distinctions might include melodic looping (such as the discussion of *Son-*

ic's menu music in chapter 2) versus the repetition of bass ostinatos (heard in some retro games like *Celeste* and *The Messenger*). At this short motivic level, such vocabulary will help in identifying processes of variation and intervallic manipulation (such as that presented in chapter 3). However, some additional vocabulary will be useful for discussing musical repetition at hierarchical levels beyond the motive. This book will employ William Caplin's basic ideas (pun intended) on phrase structure to map out some of these repetitions.[47] While this might seem like an odd choice considering how strongly Caplin's theories are grounded in the music of the classical era, his framework for phrase structure allows for basic distinctions that occur in musical styles well beyond the classical canon. Many phrases begin with a repetition of its opening motif, what Caplin terms the *basic idea*. Other phrases feature the juxtaposition of an opening motif with contrasting material, which Caplin terms an *antecedent* pairing (after typical period forms). These form part of larger structures such as sentence, period, and hybrid phrase models that give us some basic vocabulary to discuss repetition versus contrast.

Huron indicates that listeners perceive music hierarchically, and that as Western listeners we have a tendency to group phrases "into eight-measure 'periods' and sixteen-measure 'double-periods.' These groupings often . . . coincide with the boundaries between phrases."[48] He also says that "in Western classical music there exists a preference for binary beat groupings and a marked preference for binary beat subdivisions."[49] Both of these expectations will allow us to discuss irregular phrase lengths as moments of incongruence with expectations (such as the discussion of *Sonic*'s "Game Over" music in chapter 3). Phrases themselves will be organized into larger repetitions of music, which I will use letter names to represent throughout this book (such as "AB" to represent what we might commonly call binary form), with the addition of framing concepts such as introduction and outro/coda. Part of the reason I will primarily use letter names to represent these formal structures is that the music of video games draws equally from precedents in popular music and jazz in addition to art music styles; sticking to letter names to represent large-scale sections will remove the need to prioritize one genre over another. Additionally, given the lack of precision about the number of repeats in many video game contexts, forms that might be understood as classical rondo, pop's verse/chorus form, or a jazz form such as thirty-two-bar form will thus reduce down to an AB structure with possible transition sections. In many cases, the exact form is not essential to an understanding of the music in its context; rather, the moment of change to a new musical section is the more important structural feature given its frequent correspondence to a change in the game state.

4. Timbre and Instrumentation Play a Significant Role   Finally, it is worth mentioning that in this text I will place a stronger emphasis on timbre and instrumentation than might typically be expected in music analysis. In the Western art music approach to musical analysis, instrumentation is often at best a footnote, with the main analytical focus relying on harmony and rhythm (and changes in instrumentation seen merely as rearrangements that do not impact the overall musical structure). In gaming, however, I believe that players perceive instrument and timbre as essential components in all three forms of authenticity. Instrumentation, for instance, strongly relates to player perceptions of intertextuality (particularly of tropes from previous media experiences, discussed in chapter 2) and of the various geographies associated with gameworld (chapter 1). Thematic associations are more clearly perceived through similar instrumentation, and changes of instrumentation suggest changes in the narrative state (chapter 3). Or alternately, the use of 16-bit sounds is one of the key signifiers of retro game sound (chapter 4). Discussions of timbre might also speak to the artificiality or naturalness of the sound as perceived by the player, providing a context for the discussion of objective and constructive authenticity. In this book, timbre (either identified via its intended modelled instrument or as a duality between digital synthesized timbres and more realistic sampled instrumental sounds) will thus be a significant component of each analysis.

FINAL NOTES BEFORE WE GET STARTED

An analysis of the game's music alone is needlessly one-dimensional given that the interaction between audio, visuals, and narrative is a vital part of most games. Because such a multifaceted approach can be difficult to depict through the two-dimensional format of print, I highly encourage you to watch (or play) the various scenes from the game examined in this study. YouTube links will be included when possible, but given the ever-changing application of copyright (Nintendo, for instance, has shifted their policies over time between supporting player video uploads through their partner program to banning them completely in later instances) and the focus on interactivity, gameplay is the best mode of engaging with this content. As players and readers, our understanding of being "in the game," with its rules, conventions, and other components of the gameworld, impacts musical experience in ways that are difficult to articulate through print. In this book's conclusion, I will explore the full implications of this integration: that gameplay (including its music) exists as an *oeuvre* or work of art, and how authenticity is perceived in this larger context.

NOTES

1. For those interested in this debate, Philip Ewell's article "Music Theory and the White Racial Frame," *Music Theory Online* 26, no. 2 (September 2020), can be found at https://mtosmt.org/issues/mto.20.26.2/mto.20.26.2.ewell.html. An overview of the reaction to Ewell's work can be found in Colleen Flaherty, "Whose Music Theory?" *Inside Higher Ed*, August 7, 2020, https://www.insidehighered.com/news/2020/08/07/music-theory-journal-criticized-symposium-supposed-white-supremacist-theorist.

2. For an excellent discussion of this issue, please see Yussef Cole, "Cuphead and the Racist Spectre of Fleischer Animation," *Unwinnable*, November 10, 2017, https://unwinnable.com/2017/11/10/cuphead-and-the-racist-spectre-of-fleischer-animation/.

3. Karen M. Cook, "'The Things I Do for Lust . . .': Humor and Subversion in *The Bard's Tale*," in *Music in the Role-Playing Game*, ed. William Gibbons and Steven Reale (New York: Routledge, 2019), 26.

4. See, for example, Tim Summers, "Analysing Video Game Music: Sources, Methods and a Case Study," in *Ludomusicology: Approaches to Video Game Music*, ed. Michiel Kamp, Tim Summers, and Mark Sweeney (Sheffield, UK: Equinox Publishing, Ltd., 2016), 9, and Karen M. Cook, "Beyond (the) *Halo*: Chant in Video Games," in *Studies in Medievalism*, ed. Karl Fugelso, Vol. XXVII (Suffolk, UK: Boydell & Brewer, 2018), 185.

5. Isabella van Elferen, "Analysing Game Musical Immersion: The ALI Model," in *Ludomusicology: Approaches to Video Game Music*, ed. Michiel Kamp, Tim Summers, and Mark Sweeney (Sheffield, UK: Equinox Publishing, Ltd., 2016), 33–34.

6. Ibid., 35.

7. Tim Summers, *Understanding Video Game Music* (Cambridge: Cambridge University Press, 2016), 40.

8. William Gibbons, *Unlimited Replays: Video Games and Classical Music* (New York: Oxford University Press, 2018), 29.

9. Ibid., 34.

10. Jesper Juul, *Handmade Pixels: Independent Video Games and the Quest for Authenticity* (Cambridge, MA: MIT Press, 2019), 9.

11. Richard A. Peterson, *Creating Country Music: Fabricating Authenticity* (Chicago: University of Chicago Press, 1997), 199–201, as cited in Juul, *Handmade Pixels*, 48.

12. Elizabeth Randall Upton, "Coconut Clops and Motorcycle Fanfare: What Sounds Medieval?" *Sounding Out!* (blog), September 19, 2016, https://soundstudiesblog.com/2016/09/19/coconut-clops-and-motorcycle-fanfare-what-sounds-medieval/.

13. Richard Stevens and Dave Raybould, "The Reality Paradox: Authenticity, Fidelity and the Real in *Battlefield 4*," *The Soundtrack* 8, nos. 1–2 (2015): 64.

14. Ibid., 60.

15. Felix Zimmermann, "Approaching the Authenticities of Late Modernity," in *History in Games: Contingencies of an Authentic Past*, eds. Martin Lorber and Felix Zimmermann (Bielefeld: Transcript Verlag, 2020), 15.

16. Michał Mochocki, "Heritage Sites and Video Games: Questions of Authenticity and Immersion," *Games and Culture* 16, no. 8 (2021): 954.

17. Ibid., 13.

18. Karen M. Cook, "Music, History, and Progress in *Sid Meier's Civilization IV*," in *Music in Video Games: Studying Play*, eds. Neil Lerner, K. J. Donnelly, and William Gibbons (New York: Routledge, 2014), 175.

19. Note that throughout this book, to avoid confusion I will use lowercase *fantasy* to refer to the opposite of reality, and uppercase *Fantasy* to refer to the genre. Direct quotes, however, will use the case of the original source material.

20. Isabella van Elferen, "¡Un Forastero! Issues of Virtuality and Diegesis in Videogame Music," *Music and the Moving Image* 4, no. 2 (2011): 30.

21. Ibid., 32.

22. Summers, *Understanding Video Game Music*, 86.

23. Summers, "Analysing Video Game Music," 8.

24. Elizabeth Medina-Gray, "Meaningful Modular Combinations: Simultaneous Harp and Environmental Music in Two *Legend of Zelda* Games," in *Music in Video Games: Studying Play*, eds. K. J. Donnelly, William Gibbons, and Neil Lerner (New York: Routledge, 2014), 104–121; "Modularity in Video Game Music," in *Ludomusicology: Approaches to Video Game Music*, eds. Michiel Kamp, Tim Summers, and Mark Sweeney (Sheffield, UK: Equinox Publishing, Ltd., 2016), 53–72; "Musical Dreams and Nightmares: An Analysis of *Flower*," in *The Routledge Companion to Screen Music and Sound*, eds. Miguel Mera, Ronald Sadoff, and Ben Winters (New York: Routledge, 2017), 562–576; and "Analyzing Modular Smoothness in Video Game Music," *Music Theory Online* 25, no. 3 (2019).

25. Karen Collins, *Playing with Sound: A Theory of Interacting with Sound and Music in Video Games* (Cambridge, MA: MIT Press, 2013).

26. Sean Atkinson, "Soaring Through the Sky: Topics and Tropes in Video Game Music," *Music Theory Online* 25, no. 2 (2019): 12–17, https://mtosmt.org/issues/mto.19.25.2/mto.19.25.2.atkinson.pdf.

27. Ibid., 18.

28. Ibid., 19.

29. Gibbons, *Unlimited Replays*, 21–22.

30. Steven Reale, "Variations on a Theme by a Rogue A.I.: Music, Gameplay, and Storytelling in *Portal 2*," Parts 1 and 2, *SMT-V* 2, no. 2 (July 2016), http://www.smt-v.org/archives/volume2.html#variations-on-a-theme-by-a-rogue-ai-music-gameplay-and-storytelling-in-portal-2-part-1-of-2.

31. Ibid.

32. Ibid.

33. Steven Reale, "Analytical Traditions and Game Music: *Super Mario Galaxy* as a Case Study," in *The Cambridge Companion to Video Game Music*, eds. Melanie Fritsch and Tim Summers (Cambridge: Cambridge University Press, 2021), 218.

34. van Elferen, "¡Un Forastero!," 33.

35. Iain Hart, "Semiotics in Game Music," in *The Cambridge Companion to Video Game Music*, eds. Melanie Fritsch and Tim Summers (Cambridge: Cambridge University Press, 2021), 236–37.

36. For a more detailed discussion of this subject, see Tim Summers, *Understanding Video Game Music*, chapter 3, as well as his article "Epic Texturing in the

First-Person Shooter: The Aesthetics of Video Game Music," *The Soundtrack* 5, no. 2 (2012): 131–151.

37. Michel Chion, *Audio-Vision: Sound on Screen*, trans. Claudia Gorbman (New York: Columbia University Press, 1994), 25–28. Chion also gives a third category of listening, *reduced listening*, which focuses on the properties of the sound itself. Given our focus on the interactivity between sound, visuals, and gameplay, this third category will not be discussed here.

38. John Roeder, "Introduction," in *Analytical and Cross-Cultural Studies in World Music*, eds. Michael Tenzer and John Roeder (Oxford: Oxford University Press, 2011), 11.

39. Ibid., 11–12.

40. David Huron, "A Comparison of Average Pitch Height and Interval Size in Major- and Minor-Key Themes: Evidence Consistent with Affect-Related Pitch Prosody," *Empirical Musicology Review* 3, no. 2 (2008): 59.

41. Joe Gilliver, "Composing Music for Video Games—Key and Tempo," *Game Developer*, last modified January 10, 2014, https://www.gamedeveloper.com/audio/composing-music-for-video-games---key-tempo.

42. Huron, "Comparison of Average Pitch Height and Interval Size," 59.

43. I am consciously including the Ionian and Aeolian modes in this list despite their mappings on the major and minor scales, respectively, as they will provide a useful as a point of comparison in those chapters.

44. David Huron, *Sweet Anticipation: Music and the Psychology of Expectation* (Cambridge, MA: MIT Press, 2006), 74.

45. Huron, *Sweet Anticipation*, 154–156.

46. For instance, Sara Bowden uses the term in their article "Not Suitable for the Easily Disturbed: Sonic Nonlinearity and Disruptive Horror in *Doki Doki Literature Club*," *The Soundtrack* 11, no. 1 (2020): 7–22. However, among gamers and game designers, the term sometimes refers to adaptive or dynamic audio; that is, sound design that changes based on game events and/or player actions.

47. For an overview of these, please see William Caplin, *Classical Form: A Theory of Formal Functions for the Instrumental Music of Haydn, Mozart, and Beethoven* (New York: Oxford University Press, 1998).

48. Huron, *Sweet Anticipation*, 179.

49. Huron, *Sweet Anticipation*, 195.

One

Gameworld

OVERVIEW

This displacement of the documentary function of games—those ostensibly based on "actual" events—toward a Hollywood or pop-media measure of their realism suggests that something more is going on than the straightforward re-enactment of history. In all of these examples . . . video games operate with a clear—and a clearly mediated—relationship to the past.[1]

Zach Whalen and Laurie Taylor, in their book *Playing the Past: History and Nostalgia in Video Games*, astutely observe that "the past" in gaming is somewhat of a construct.[2] Mediation, in their description, is a substantial component of the player experience. Regardless of whether a game aims to replicate historical events or not, players' perceptions are colored by their own knowledge, (mis)conceptions, and expectations for the established gameworld. References to the past in video games can thus be viewed through the lens of constructive authenticity (and in fact, mediation can be understood as a form of constructive authenticity) in addition to the objective authenticity of the original historical reference.

This chapter will use two genres of games as the lens for examining player mediation from a music-analysis perspective. In the first section, I will look at worldbuilding and elements of historicism through an in-depth examination of sea shanties in *Assassin's Creed: Odyssey*. Modes, rhythm, and accompaniment all serve to suggest the Ancient Greek setting of the game in ways that do not necessarily correspond to historical practice. Do players notice this difference, and if so, what is the impact on their understanding of the narrative? In the second portion of this chapter, I will shift to the survival horror genre to examine in depth how intimately sound is tied to the player's understanding of their environment—that is, gameworld. The term *gameworld* is often used by players but not always concretely identified in the scholarship, so let us begin by defining this term and contextualizing it within current scholarship.

Gameworld

What is a gameworld? Surprisingly, scholars tend to avoid this term, making it somewhat difficult to pin down. Katie Salen and Eric Zimmerman, for example, use Rune Klevjer's term *fictive worlds* in their examination of game structure and design, *Rules of Play*, and recognize the reciprocal relationship between fictive worlds and story events: "The story events of shooting, maneuvering, destruction, and survival gain narrative meaning from the larger fictive world of the game, even as they simultaneously help to define that fictive world."[3] Scholars Tison Pugh and Angela Weisl, in their examination of modern-day medievalisms, give a definition without actually providing the term itself: "produc[ing] its own world within a system of meanings that overrides the rules of realism while still remaining believable . . . creating an authentic atmosphere that allows the game's story to be played out within its constructed framework."[4] Isabella van Elferen discusses this constructed framework in terms of *virtuality* and *game spaces*, and argues that music "dismantles the borders between virtual and real spaces, undoing the computer interface and replacing it with the 3-D interface of aural imagination."[5] Some commonalities arise from these various perspectives. These fictional universes are self-established and self-referential, linked to player imagination, and as a result tie strongly to the player's sense of (constructive) authenticity. Whether we call them constructed frameworks, fictive worlds, or game spaces, gameworlds have a strong link to the concept of the *magic circle*, the player's mental conceptualization of gameplay. Ernest Adams thus defines gameworld as "an artificial universe, an imaginary place in which the events of the game occur. When the player enters the magic circle and pretends to be somewhere else, the game world is the place she pretends to be."[6]

Gameworlds present elements of the fictional universe both within and outside of the player-observed narrative frame: mythology, history, geography, and cultural elements such as costume, architecture, music, and more. Games achieve this through not only narration and direct story presentation, but also through a variety of visual elements and (more importantly for this discussion) sound. However, sound and music are often an afterthought among scholars discussing gameworld. In her 2013 book *Playing with Sound*, Karen Collins cites scholar Michael Nitsche, who states that "the screen remains an important layer as it is mainly through the screen that the game worlds can unfold and become accessible to today's player."[7] Collins, however, refutes this statement, arguing that sound is an important means of constructing and perceiving virtual spaces such as gameworld. Van Elferen further argues that music "expands the gaming magic circle into the real-life surroundings of the player; in this way it underlines the potentially uncanny dimension of videogaming" in its blurring of boundaries between real life and the virtual universe.⁸ We can thus consider a binary opposition between gameworld and real life, but with some instances crossing the border between the two and tying into scholarship on the magic circle.

How does sound help to build gameworld? As Collins and van Elferen allude to, sound in games can play a role in creating immersion by blurring distinctions between the gameworld and the player's real-world environment. William Cheng delves into this question in detail in his examination of the horror game *Silent Hill*:

> Even more disturbing than the muddling of diegetic and non-diegetic noises is how these sounds cross from the game's virtual world into the real world inhabited by the player. Buzzes, rumbles, door slams, and other noises bleed with ease from the gameworld into a player's own environment precisely because they sound like everyday racket. We may occasionally be duped into hearing these in-game sounds as if they're coming from real-world sources—from our living rooms, from upstairs, from right outside our windows. Liable to fool us in this regard are not the game's outrageously loud, dissonant, or repetitive noise samples, but rather the little mundane sounds that pop up now and then: a creaking floorboard here, a muted thump there, a generic beep from nowhere. The sheer density of this industrial audio is enough to create the impression of surround-sound; that is, noises in the game—even when piping out of a television's speakers—can so extensively saturate a player's physical space that they might sound as if they are invading from all directions.⁹

Cheng's example, despite (or perhaps because of) the blurring of the magic-circle boundary, acts to establish the gameworld: a horror-themed world that uncannily replicates the real world and uses unpredictability to establish mood, but in other cases transforms sound that possesses little to no semantic meaning in the real world into meaningful sounds that cue the player towards particular in-game actions. And in borrowing from the real world, the game achieves a particular type of authenticity for the player where players expect the game to conform to their expectations of the real world, especially in its visuals and sound, despite the fantastic elements of horror superimposed on this framework.

Throughout the discussion in this chapter, I will highlight several ways that sound and music establish gameworld—an important precursor to discussing the perception of authenticity—through three functional roles. While these roles are inspired by previous scholarship in game sound by Zach Whalen and in film sound by David Neumeyer and James Buhler,¹⁰ I would like to expand them to focus more specifically on video games' more interactive role and on differences particular to gameworld itself. These functional roles for game sound include:

1. *Establishing time and place,* which can include elements to create the historical setting (for example, the use of particular regional or historical musical instruments) or to suggest the size or layout of a particular game environment (such as the addition of reverb to indicate larger spaces). This is a form of texturing, as defined in the introduction of this book—filling in detail of the gameworld through both current and previous knowledge.
2. *Creating narrative inflection,* which can include overtly expository narrative elements, such as dialogue; elements, such as voice timbre, that suggest the particular emotional states of characters without stating it overtly; or components, such as music, that might provide an emotional subtext after mediation by the player.
3. *Cueing;* that is, sounds that indicate that a player should take a particular action (such as a sound effect that increases in volume as a player nears a game objective), the current state of gameplay (for example, a low-health warning beep), or success in the attempted action (such as the *cha-ching* sound effect often associated with picking up treasure or coins). If such sounds motivate future actions, they may also be understood as evoking, as described in the introduction of this text.

I believe that these three elements of gameworld sound cannot be untwined from players' perceptions of immersion, authenticity, and successful gameplay realization. I will begin this chapter's case studies by delving into the first function, *establishing time and place.*

ESTABLISHING TIME AND PLACE

Historical Authenticity versus Gameplay Authenticity in the Fictional Worlds of Ancient Greece

The *Assassin's Creed* series (and its publisher, Ubisoft) is well known for its devotion to historical research. As one source puts it:

> Regardless of time and location, *Assassin's Creed* provides its players with a richly reconstructed past world: building materials and styles are period accurate, city layouts and landscapes conform to what is known about their original geography, other material culture mostly adheres to the style and technology of the day, people tend to be dressed and act appropriately, and scenes of daily life fit within scholarly expectations. Added to this general eye for detail is a penchant for historical drama: famous landmarks, events, and persons all feature prominently in the games and have often been carefully researched and precisely represented.[11]

Marketing press releases, interviews, and media articles surrounding the release of each game in the *Assassin's Creed* series additionally position the game within an authentic historical setting, a claim often reinforced through interviews with historians, linguists, and other scholars from well-known academic institutions.[12] Players who consume this media can't help but be influenced by the message sent: *Assassin's Creed*, despite being a fictional story, re-creates historical environments meticulously and as accurately as possible.

But is this actually the case? I recently had the pleasure of playing *Assassin's Creed: Odyssey*, an expansive, open-world adventure game set in Greece in the fifth century BC. The two opening cut scenes approach the sound of gameplay in two very different ways. The first, a battle scene (to be discussed in depth later in this chapter), features a musical scoring typical of the so-called "epic" game style. After this opening battle scene, the narrative jumps forward in time to a vastly different setting. The player selects their choice of male or female character and then is launched into a new cut scene of the Greek countryside (figure 1.1).[13] A number of elements serve to establish the historical time and place of the game in this scene, including a literal eagle-eye view of the player's starting location, soaring through the air to present the mountainous, rocky terrain, larger-than-life statues of Greek gods, and a panoramic view of the sea. The statues and ancient temples included in this scene provide a striking signifier of Ancient Greece, but also present a curious anachronism as this signifier is more in line with stereotypes of modern Greece. The ancient temples of today would presumably have been the current-day temples in 430 BCE, and such ruins would neither occur that frequently in the landscape nor be that weathered and old in that era!

Figure 1.1. Screenshot from the opening bird's-eye-view cut scene, *Assassin's Creed: Odyssey*. **Source**: Reproduced by permission of Ubisoft Entertainment SA.

Such signifiers also exist in the soundscape of the scene. The cry of an eagle, a prominent feature throughout the *Assassin's Creed* series, is heard multiple times in the course of the one-minute panorama view; a slight sound of wind can be heard in the background; but most prominently a female voice is heard singing, accompanied by bouzouki, a traditional Greek plucked-string instrument. The elements in combination suggest Ancient Greece to the player . . . or do they?

Historicism in Game Sound

In his book *Unlimited Replays*, William Gibbons identifies that game designers, "rather than conforming to historical fact . . . often will appeal to what players *think* they know about music history. In other words, they find a sweet spot between, on the one hand, conforming slavishly to historical fact and, on the other, alienating players by stretching reality too far—an intriguing interplay."[14] His study references this effect in connection to the use of classical music in video games, but such practice is common in games referencing a wide array of historical content and can even create new intertextual frames of reference that are brought forward into players' future gaming experiences.

The *Assassin's Creed* series is one example of this phenomenon. With each game situated in a particular historical time and place, the music helps to create authenticity for the player—but the term *authenticity* in this case refers to the player's expected gaming experience (constructive authenticity) and not necessarily to the historical time and place being emulated (objective authenticity). For *Assassin's Creed: Odyssey*, the expected gameworld includes a myriad of references to Ancient Greek mythology, geography, culture, histories, poetry, and more. As Aris Politopoulos et al. describe, "despite historical inaccuracies, the game provides visual and virtual authenticity. . . . Despite its compressed size and the artistic liberties taken with the layout of neighborhoods, a player feels as if he or she is exploring the actual city."[15] While there is certainly a strong attempt towards realism in the game design, this authenticity has its limits since the main focus in on creating an enjoyable gaming experience. To give just one example, most of the game's dialogue is delivered in English (or the player's chosen localized language) rather than in Ancient Greek so that players can more immersively experience the dialogue in real time.

The music of the game was designed as a deliberate compromise between historicism and an immersive player experience by its composers: collaborative duo The Flight (consisting of Joe Henson and Alexis Smith), Mike Georgiades, and Giannis Georgantelis, along with further assistance from Kalia Lyraki, Emma Rohan, and Panagiotis Stefos. For example, The Flight, in a 2018 interview for Ubisoft, explained that they were more concerned about creating an overall feel for the game rather than aiming for historical reproduction:

> Our first step on this [compositional] path with *Odyssey* was researching the historical and geographical setting, and this gave us an "in" as to how we wanted it to sound: Ancient Greek in flavor and feeling whilst still being a contemporary score. . . . We listened to a lot of the reproductions of Ancient Greek music and drew a lot of inspiration from them, especially in terms of harmony and instrumentation. Of course, we weren't trying to re-create this in our score, as that was the job of the diegetic music, so it was more for vibe and feeling.[16]

While they cite their attempt to imitate instruments that might have been available at the time, they also acknowledge that the end product used modern-day instruments to achieve those sounds:

> We did a lot of research into what instruments might have been around at the time, and then tried to find their closest relatives available today. We bought a hammered Dulcimer, and various Lyres, bouzoukis and panpipes. One of the instruments we were keen to involve was the Diaulos; we had seen some videos of people playing them online and loved the piercing sound, quite unlike anything we had heard before! Unfortunately, it proved impossible to get hold of one, but we used bagpipe chanters and a blown reed to get an approximation.[17]

Nevertheless, despite a focus on feel rather than historical detail, players often cite the game for its perceived historical accuracy. One viewer's comment on the soundtrack's YouTube page illustrates this well, describing the game's music as:

> Amazing quality. . . . It would have been interesting if it was mentioned which pieces/piece exactly have been composed by Mr Georgiades :-D What is worth of admiration, and why—as a Greek—respect Ubisoft for, is that many pieces remind me of contemporary greek music. Makes me feel like home.[18]

The scene "feels like home" for the modern-day Greek player not because of its connection to the time of Ancient Greece, but rather based on its association with their modern-day understanding of place and their expectations of game music styles. This is perhaps best illustrated by the instrumentation. The bouzouki used to accompany the female voice in "Odyssey," the song featured during the introduction to the game's main character, is a more recent arrival to Greece, imported by immigrants from Turkey in the early 1900s. More emblematic of modern Greek folk styles than of Ancient Greece, Stathis Gauntlett describes that the bouzouki and its associated Rebetika folk genre have been anachronistically assigned a direct lineage from Ancient Greek music for nationalist and political reasons. According to Gauntlett, "the process by which music, dance and instruments that were once dismissed as Turkish and embarrassingly crude entered the showcase of diachronic Greek culture is as complex as it is remarkable."[19]

An analysis of several musical examples from *Assassin's Creed: Odyssey* will demonstrate how the composers evoke "Ancient Greek" for their players. The analyses will compare mode, meter, and texture to isolate features of this (pseudo) Greek style, and will further demonstrate how *Hades*, a game that aims for *less* historical accuracy, nevertheless incorporates similar musical features. These musical cues both evoke players' expectations of authenticity and actively communicate information significant to gameplay.

Analysis

The music of *Assassin's Creed: Odyssey* in many ways differs significantly from its predecessors—not surprising given a shift in emphasis in the plot. Over the last fourteen years fans and critics alike have panned the series' emphasis on two timelines, one ancient and one modern, as confusing and artificial. *Odyssey*, while still incorporating this dual timeline, shifts the ratio of ancient to modern significantly away from the modern and towards the historic, and the music reflects this difference.[20] While the earliest games in the *Assassin's Creed* series folded historical elements into a soundscape that more strongly emphasized digital distortion and an electronic dance music style, *Odyssey*'s style leans towards a mainly acoustic sound emphasizing historical styles, and more modern elements such as symphonic instruments are relegated to a supporting role. One such example of this can be heard in the track "Kephallonia Island," featuring santur and psaltery as the lead melody and prominent drone instruments with additional finger cymbals and tambourine. As the song progresses, violins (a decidedly non-Greek instrument) enter in a supporting role to double the drone tone and create a thicker density of sound.[21]

Diegetic and non-diegetic music share some rhythmic and modal features in *Odyssey* but differ in their instrumentation. While the non-diegetic music within the game often includes key elements of epic game music style, including prominent low-register strings and brass, a rapid, continuous rhythmic pulse, and a strong emphasis on low-register percussion, the diegetic music is scored mostly with voice.[22] The main sources of in-game diegetic music are a cappella sea shanties performed by the protagonist's sailing crew, songs (typically with voice and plucked strings) performed by non-player character (henceforth "NPC") musicians scattered throughout the gameworld, as well as the previously discussed track "Odyssey," a semi-diegetic song that transitions from non-diegetic to diegetic. The sea shanties are divided into English- and Greek-text songs. Since the Greek sea shanties form a coherent group of songs composed by a single composer, Giannis Georgantelis, for a single function within the game—providing some local color to the boat voyages—they will provide a useful subset of repertoire to illustrate some of the recurring features of rhythm and pitch in *Odyssey*'s music.

Figure 1.2 gives the rhythms and pitches for the sea shanty "Poseidon, God of the Sea."[23] The piece is in D Phrygian mode,[24] established through repetition of D in the first four bars as well as the drone voice in bars 9–16. The tonal focus shifts to G at bar 17, with accidentals C♯ and F♯ introducing some chromaticism in bars 22–23. Added E-naturals in the section starting at bar 25 suggest a modulation to D Aeolian (that is, natural minor).

Figure 1.2. "Poseidon, God of the Sea" with Latin alphabet transliteration. *Source:* Composed by Giannis Georgantelis. Transcription and analysis by the author. Reproduced by permission of Ubisoft Entertainment SA and Hal Leonard, Inc.

Rhythmically, while I have transcribed this work in 3/4 time due to its repetition of four-bar groupings, the word emphasis and syllabification suggest a different accent structure than would be expected in this meter. As shown by the commas added to the figure (and audibly articulated in the recording), the word grouping creates an accent on the second eighth note of bars 2 and 4 and analogous locations. This disrupts the listener's sense of triple meter, instead creating a sense of shifting downbeat throughout this section. The off-beat entry of the third voice in bar 9 produces a similar accentuation. The final section (beginning in bar 25) repeatedly changes between duple and triple subdivisions, another way of disrupting regularity in this song.

This is surprisingly consistent with what scholars currently know about Ancient Greek musical practice. As M. L. West states in his book *Ancient Greek Music*, in the fifth century BC "[for] the Greek composer of vocal music . . . his rhythmic system was 'additive,' built up from units of fixed size, as opposed to the 'divisive' principle of Western music in which the constant is a measure of time (a bar) that may be divided into fractions of many different sizes."[25] West also indicates that it was common to mix rhythmic groupings of different sizes—what we modern listeners would describe as changing meter—and consistent with the shifting word emphasis in "Poseidon." However, he also indicates that long strings of shorter-duration notes like those in bars 1–16 of "Poseidon" were avoided in Ancient Greek vocal music given the strong emphasis that word stress played in determining meter.[26] "Poseidon" clearly takes a hybrid approach, using syncopation and rhythmic irregularity to signal Ancient Greek to its players over the type of repeated eighth-note pulse commonly used in video game music to signal energy and drive.

Texture, melody, and accompaniment in the work are handled in consistent ways by Georgantelis and signal Greek-ness as well. In "Poseidon" and several of the other sea shanties, second voices often repeat a drone pitch on scale-degree 1, such as in bars 9–12. Bars 13–16 feature a repetition of this drone in the outer voices with the inner voice introducing parallel fifth movement, and bars 17–21 further emphasize this tone through a longer duration and repeated semitone motion from E♭ to D. Furthermore, the parallel perfect-fifth motion introduced in bars 13–16 returns in bars 27–28, shifting to parallel perfect-octave motion in bar 29. Tonic drones, parallel fifths, and parallel octave motion are Georgantelis's most common forms of adding additional voices in these songs and once again signal Greek-ness to the player.

Two other sea shanty examples show a similar drone and parallel perfect-fifth "harmonization." In "Song for a Young Girl" (figure 1.3), the first four bars begin by emphasizing B, thereby suggesting B Phrygian mode, but this shifts to a focus on E (suggesting E Aeolian) starting with

Figure 1.3. "Song for a Young Girl" with Latin alphabet transliteration. *Source*: Composed by Giannis Georgantelis. Transcription and analysis by the author. Reproduced by permission of Ubisoft Entertainment SA and Hal Leonard, Inc.

a transposition of the melody in bar 5 and continuing for the remainder of the song. Georgantelis ties the two tonal areas together by modifying the transposition of the melody in bars 8–9. While we expect an exact transposition to leap down by third from the end of bar 2 to the beginning of bar 3, in bar 9 the melody moves by step to land on B instead of A. The melody then finishes via a descending minor triad through scale-degrees 5-3-1 rather than bar 3 and 4's Phrygian 4-2-1 motion. Interestingly, the modified notes are those that articulate the defining feature of each mode to modern listeners, with Phrygian emphasizing the ♭2 scale-degree versus Aeolian's close ties to minor mode articulated through triadic harmony. Two-voice homophony beginning in bar 6 features an E drone that often creates the interval of a perfect fifth with the melodic tone B, such as in bars 6, 7, 9, and 19, further linking the pitch centre of the two different tonal areas.

Figure 1.4 presents another example, "Ares, God of War," in G Phrygian mode. While the first five bars are monophonic, in bar 6 the voices split into two groups. Of the total vertical (that is, harmonic) intervals in bars 6–12, the majority (five) are perfect fifths, with three additional perfect fourths (the inversion of the perfect fifth), one minor seventh, and one major third. Clearly there is an attempt to emphasize the perfect fifth not only between scale-degrees 1–5, as heard in bar 6, but also between scale-degrees 7–4 in bars 7 and 10. Its inversion, the perfect fourth, is emphasized between scale-degrees 1–4 in bars 7 and 12.

Figure 1.4. "Ares, God of War" with Latin alphabet transliteration. Composed by Giannis Georgantelis. Transcription and analysis by the author. Reproduced by permission of Ubisoft Entertainment SA and Hal Leonard, Inc.

Does this resemble the harmonic practice of Ancient Greece? West gives little detail about vocal harmony in his study but does describe some examples of how voice was accompanied instrumentally. In studying a fragment from Euripides's *Orestes*, West suggests that the notation implies drone tones and the use of concords (that is, the perfect consonances: the perfect fourth, fifth, and octave).[27] Drones are a particularly common choice in modern media to represent the musically ancient, so in this case the history and player expectation may coincide.

However, despite this one correspondence, other elements of pitch are wildly off from the historical reality. Table 1.1 lists the major features of mode, meter, and voice interaction within *Odyssey*'s Greek-language sea shanties, summarizing observations about voice interaction and drone, but also more explicitly detailing mode, emphasized tones (focal pitches), and meter. The modes used are surprisingly modern: of the ten songs listed, four occur in (modern) Phrygian, two are in major/Ionian, five are in Aeolian or another form of the minor mode, and the final song is in the mode sometimes termed "Phrygian major" (that is, Phrygian with a raised scale-degree 3). All the sea shanties are performed in modern-day equal-tempered tuning, a focus on diatonic scale organization predominates, and microtonal scales are notably absent. This is contrary to what we might expect from Ancient Greek music. While diatonic scales were sometimes used in that time period, the enharmonic and chromatic scale types were more common.[28] These enharmonic and chromatic scales were constructed from two consecutive tetrachords, each spanning the interval of a perfect fourth. As shown by West in figure 1.4, while the diatonic scale type was structured via tones and semitones, the enharmonic and chromatic varieties were less evenly spaced, instead clustering three pitches at the bottom of each tetrachord followed by a larger gap to complete the perfect-fourth-framing interval of the tetrachord. These pitches were approximately quarter tones in the enharmonic scale and semitones in the chromatic scale.

Table 1.1. Musical properties of the Greek-language sea shanties in *Assassin's Creed: Odyssey. Source:* Table by the author.

Song	Mode	Meter (as transcribed)	Voice Interactions*
Ares, God of War	G Phrygian Emphasized tones: G, D, C	Simple triple (3/4)	Homophony and polyphony Vertical intervals: P5 (x5), P4 (x3), m7 (x1), M3 (x1)
Bacchus Teaches Me to Dance	B♭ Ionian (major), modulating to B♭ Aeolian (minor) or Dorian Emphasized tones: B♭, F, D	Simple quadruple (4/4)	Call-and-response, monophony
Muse of the Forest	G Aeolian (minor) Emphasized tones: G, F	Simple quadruple (4/4)	Call-and-response
Poseidon, God of the Sea	D Phrygian, but B section adds C♯, F♯ suggesting G minor with a raised scale-degree 4 Emphasized tones: D, A, G (esp. in B section)	Simple triple (3/4), but implies a change between 3/4 and 6/8 based on long-note duration and there are also significant off-beat accents that shift the pulse an eighth beat later throughout	Melody over/under drone; parallel 5th motion
Song for a Young Girl	B Phrygian shifting to E Aeolian Emphasized tones: B, E	Simple quadruple (4/4)	Melody over drone, monophony
Song to Bacchus	C, D, E♭, F, G, A, B = Jazz melodic minor; pitch center shifts from G in the first section to C in the second section Emphasized tones: G, D, F in first section; C, G in second section	Simple quadruple (4/4)	Call-and-response, melody over drone

(continued)

Table 1.1. *(continued)*

Song	Mode	Meter (as transcribed)	Voice Interactions*
The Black Earth Drinks	D minor in both harmonic and melodic forms Emphasized tones: D, A	Compound duple (6/8)	Monophonic
The Lost Shield	D Ionian (major) Emphasized tones: D, A, E	Simple quadruple (4/4) but sounds like an irregular meter because of positioning of longer durations and changing phrase lengths	Monophonic
Through the Storm	E Phrygian Emphasized tones: E, F, G	Compound duple (6/8), with compound triple (9/8) in the final bar of each phrase	Drones on E and D
When I Drink	D Phrygian with raised scale-degree 3 (Phrygian major); sometimes adds a raised leading tone Emphasized tones: D, A, C	Simple quadruple (4/4) but sounds like an irregular meter due to positioning of longer durations on beat 3 and word syllable divisions	Monophonic

*Note that octave doublings are not explicitly indicated in this table since the members of the performing chorus sing the given melody in their most comfortable register regardless of notation.

The scales and tuning used in *Odyssey* are not, however, inconsistent with the practice of modern Greek folk music. Nikos Ordoulidis, in his article "The Greek Popular Modes," explains that modern Greek "popular" (or *laikó*) modes include "elements [that] emanate from the maqam system, the Byzantine system and the western tonal system and harmony," but that over the last century the microtonal structure of the maqams has been replaced with a Western-based tuning system partly because of the prominence of the fretted bouzouki (with frets positioned according to the equal-tempered tuning system) in Greek folk music performance.[29]

Ptolemy's categories are as follows. Besides giving the intervals in cents I show what each tetrachord looks like when expressed in modern notes with appropriate modifiers.

enharmonic	38	73	387	e e↑⁴⁰ f↑¹⁰ a
soft chromatic	63	120	315	e f↓⁴⁰ f♯↓¹⁵ a
tense chromatic	81	150	267	e f↓²⁰ f♯↑³⁰ a
soft diatonic	84	183	231	e f↓¹⁵ g↓³⁰ a
tonic diatonic	63	231	204	e f↓⁴⁰ g a
tense diatonic	111	204	183	e f↑¹⁵ g↑¹⁵ a
even diatonic	150	165	183	e f↑⁵⁰ g↑¹⁵ a

Figure 1.5. Tetrachord construction in Ancient Greek scales and variants, with numeric values indicating cents. *Source*: M. L. West, *Ancient Greek Music* (Oxford: Clarendon Press, 1992), 170.

Ordoulidis explains that, as a result, Greek musicians' conceptualization of scales has shifted in the last century away from a tetrachordal construction towards the diatonic system favoured in Western European art music:

> Another problematic issue is the fact that Greek musicians think of the dhrómi as being scales of eight notes, that is, octachords. They teach them in this way and they also communicate on the music stands in this way. Having a look at the few books published by bouzouki players verifies this problematic point. All the dhrómi are presented as being scales. However, the main element of the maqam system is that it emphasizes the utilization of the tetrachord and the pentachord rather than the octachord.... The importance of the tetrachord and pentachord is true for Byzantine music as well. Obviously, the way Greek musicians treat the dhrómi reveals a tendency towards western musical thinking.[30]

His article explains that triadic harmony is common in modern folk music performance, and he even presents typical triadic harmonizations for common major/minor scales and modes. While English-language academic resources on Greek folk scales are scarce, amateur bouzouki learning materials created and distributed by Greek musicians provide some insight here. Scales are typically notated with Western notation, are in octave groupings (what Ordoulidis calls "octachord" formation in the quotation above), are harmonized with triadic harmony, and often duplicate the various diatonic modes or modes of minor scale forms.[31] There is no indication of any pitches outside of the chromatic scale, and the underlying theoretical structure is taught with a focus on diatonicism (although microtonal ornamentation is heard in performance).

Authenticity

In his interview with *The Flight*, Hektor Apostolopoulos acutely observes:

> The weird thing about Ancient Greek music is that practically none of it survives. We might have a grasp of instruments that were predominantly used in the era, thanks to various artistic depictions, but we don't really, really know much about the music itself. If we Greeks were pressed to answer whether any one of the flourishes used in your work is "Ancient Greek" enough, we wouldn't really know. At the same time, the lack of information gives you creative leeway, enough room to try and capture a music feel that feels right in the end and, of course, can be built upon and used in conjunction with modern structure and cinematic sensibilities.[32]

In my view, this is the most insightful observation of the entire interview. Since little notated music survives in the historical record, an attempt to re-create Ancient Greek music is more guesswork than anything else.[33] Players, the large majority of whom are not familiar with existing scholarship on Ancient Greek music, don't really know whether the musical re-creations are accurate or not. Lacking that frame of reference, players will pull from their knowledge of other media—their "cinematic sensibilities" (or Hollywood, as Tim Summers describes).[34] Summers explores this subject in depth in chapter 4 of *Understanding Video Game Music*, stating that "music may deploy 'signs of the real' to imply realism and help construct the game world, rather than using music (or musical absence) that is closer to the actual world sonic reality."[35] Authenticity as understood by gamers thus does not mean reality (objective authenticity), but instead is understood as that which conforms closest to the expected game experience and is in the eye of the beholder (constructive authenticity)—whether that be the player, the game designer, the composer, or a combination of all three—rather than being a fixed historical fact. In *Odyssey*, players pull their expectations from their previous media experiences depicting the Ancient Greek and classical worlds, often conflating Greek and Roman cultures, but also from their previous experience with the *Assassin's Creed* games and the media hype surrounding them. And as Summers aptly observes, this shapes how players understand history as well:

> Intentionally or otherwise, in drawing on actual world history to create a virtual world, a certain amount of informal education takes place, whereby the virtual world implicitly comments on the actual world. This is particularly significant where the divergences between game world and the actual world are ambiguous for players without prior historical knowledge. . . . Games not only musically construct the virtual worlds for gamers, but they also have the potential to partly construct their understanding of the "real" world, too.[36]

Furthermore, the musical elements that stick out to the player are those that are different than their typical gaming expectations. While violins supporting the overall harmony are likely to be heard as the default style in video game music due to their ubiquitousness, the use of bouzouki or santur provides a prominent difference in timbre to the average game soundscape. Listeners' attention is drawn to these elements of difference in comparison to their expectations. Modes, meters and rhythms, and instruments that differ from the expected gaming norms of epic orchestral sound or retro 16-bit audio are thus more likely to evoke the historical past or particular cultural references for players when supported intertextually—that is, by previous media literacy—to create the gameworld. And as a result, those signifiers of rhythm, mode, or instrumentation can still evoke such references even when used in smaller quantities or in combination with the "modern structures" Hektor Apostolopoulos identifies.

Hades *and Musical Signification*

The music from the game *Hades* is a good example of how such signifiers still evoke Ancient Greece to the player despite being combined with more modern structures. Released in 2018, *Hades* is a rogue-like dungeon crawler based on characters from Ancient Greek mythology.[37] With gameplay that focuses on quick battle sequences, the style of the music often veers towards rhythmically active rock and metal styles with heavy electric guitar and percussion.[38] Darren Korb, the game's composer, states:

> I've mentioned some music recs elsewhere, but some highlights: Louis Cole, Knower, Vulfpeck, Jeff Buckley, Pixies, Imogen Heap, Neko Case, The New Pornographers, Damone, etc. . . . For this game specifically, I wanted the writing to have a vaguely Mediterranean vibe . . . but not TOO specific, since actual ancient Greek music was pretty weird. I definitely looked at some actual ancient Greek pieces and they informed the music a bit. Mostly it was the acquisition of some instruments from that part of the world that helped in this department: Lavta, Bouzouki, Baglama are the main ones. For the rockin parts, I looked specifically at Soundgarden, Alice in Chains, Black Sabbath, Opeth, Rush, Kansas . . . and many more. . . . I wrote and performed the music on a baglama, so it has a bit of that Mediterranean vibe, but is a more modern feeling chord progression, I think.[39]

Instrumentation seems to be a key signifier of Greekness here, and Korb has aimed for a style that is vaguely Greek but not too specific; baglama, for example, is similar to the Greek bouzouki but is actually a Turkish instrument. Delving into the music a bit more deeply, several of the Greek signifiers present in *Odyssey* also appear in this soundtrack. Let's take a look at a few examples.

Throughout *Hades'* soundtrack, Phrygian mode and an emphasis on scale-degree ♭2 occurs repeatedly; tracks such as "No Escape," "The House of Hades," "From Olympus," and "Rage of the Myrmidons" are all in Phrygian mode. Other tracks such as "Lament of Orpheus" mix scale-degree ♭2 and regular scale-degree 2 to suggest borrowing from Phrygian, and one track, "Out of Tartarus," appears to be composed in Locrian, the other diatonic mode featuring scale-degree ♭2. This semitone, the presence of scale-degree ♭2, and the focus on modal melody and harmony become emblematic of this game's musical style. While modes certainly are not unheard of in rock and metal styles, this particular emphasis on ♭2 is a bit unusual and sticks out to the listener. Once again, the songs are performed in modern-day equal-tempered tuning, a focus on diatonic pitch organization predominates, and microtonal scales are notably absent.

One song that emphasizes Phrygian mode is "The Painful Way," which plays during the Tartarus and Erebus levels. Figure 1.6 gives a reduction of its melody and bass line. Centered tonally around C, the melody features a four-flat diatonic collection that suggests C Phrygian mode, strongly emphasizing semitone movement from scale-degree ♭2 (D♭) to scale-degree 1 (C) through frequent repetition, as seen in figure 1.6. In later passages not illustrated here, this melody is transposed by a fifth to produce a new semitone repetition between scale-degrees ♭6 (A♭) to 5 (G), further reinforcing the structure of the Phrygian mode.

"The Painful Way" also has similar features to those discussed in *Assassin's Creed: Odyssey*. Although the accompaniment in "The Painful Way" does not function identically to *Odyssey*'s sea shanties, the repeated emphasis on C for three out of every four beats in the bass voice suggests a similar drone-like effect. It also suggests irregular rhythms despite the 4/4 time signature of the song's transcription. The main melody begins with a repetition of the pitches (C, D♭, C, B♭, C), a turn ornament over the span of three eighth notes that suggests a triplet grouping, but then the

Figure 1.6. "The Painful Way," melody and bass in the first four bars of the song's main theme. *Source*: Composed by Darren Korb. Reduction and analysis by the author. Original music copyright Supergiant Games.

bar concludes with a duple grouping. This generates a 3+3+2 rhythm that repeats in subsequent bars, suggesting an additive rhythm built up from groupings of different size, consistent with our previous descriptions of Ancient Greek rhythm.[40] In terms of voice interaction, unlike *Odyssey*, the melody and bass in "The Painful Way" appear to relate heterophonically; that is, with the same underlying structural melody alternating between C and D♭ but with each voice ornamenting the melody differently. Although it is unclear whether this was an intentional attempt on Korb's part to make the music sound more Greek, West identifies this practice as characteristic of vocal accompaniment in Ancient Greece: "Ancient sources make it clear that when a singer was accompanied by the lyre—the usual role of the instrument—its basic function was to duplicate the sung melody. From the latter part of the fifth century on, some divergent or additional notes were played as well."[41]

Another song, "No Escape" (figure 1.7), heard in the game's main menu, also implies a 3+3+2 rhythm at times (such as in bar 4), although this is somewhat obscured through the melody's use of tied rhythms. More noticeable in this example, though, is the use of irregular rhythms and meter. Taking the smallest subdivision (the eighth note) as the unit of measurement, the melody in bars 1–3 articulates rhythmic divisions of 8+3+4+9, avoiding equal subdivisions of the bar and recurring rhythmic values. Metric changes in bars 8 and 9 add one eighth beat and subtract one quarter beat, respectively, from the previous 4/4 meter, disrupting the music's regularity and instead creating a sense of instability and additive structure for the listener. The shaker and guitar accompaniment articulates a constant eighth-note pulse, which helps to establish regularity, but in bars 2, 4, 6, and 12 and subsequent repetitions of the same content the melody and bass (and consequently the harmony) change on a weaker eighth note, a similar shift of accent to that observed in "Poseidon, God of the Sea." This suggests that both composers view rhythmic and metric irregularity as a feature of Ancient Greek style (and might explain Korb's statement that "actual ancient Greek music was pretty weird").[42]

Music History versus Establishing Time and Place

> [W]hat interests me in particular is how games play with our understanding of music history. Much like the *Assassin's Creed* games treat history—sticking *just* close enough to historical fact to fit in with our prior knowledge, yet changing details to create more engaging stories—games often rely simultaneously on our knowledge and our ignorance of music history . . . the easily identifiable—and often entirely made-up—ethnic flavor of these classical works served as useful shorthand for particular cultures or locations.[43]

Figure 1.7. "No Escape," melody and bass plus accompaniment rhythm, bars 1–17. *Source*: Composed by Darren Korb. Reduction and analysis by the author. Original music copyright Supergiant Games.

This description by William Gibbons was originally intended to address games that borrow preexisting classical music, but in my view it is also spot on for the newly composed music of *Assassin's Creed: Odyssey*. The music sticks closely enough to historical musical practices to provide audible cues of "otherness" for the player, but given players' lack of knowledge about the time period, these cues are less about historical accuracy and more about authenticity to previous media literacy or players' current lived experiences. YouTube comments by modern Greeks may indicate a feeling of familiarity and home with the game's music, of music that "feels right," but this is achieved through comparison to familiar modern-day Greek folk and popular music practices rather than on the theoretical constructs outlined in the historical documentation.

Is that a bad thing? Yes and no. Summers identified that "music contributes to the 'cueing' of the virtual world of the game, replying on known musical signifiers to contextualize the game mechanic," but he also argues that such an understanding of history in-game can represent history in misleading ways.[44] I would suggest this blurring of boundaries encourages game immersion as it removes the obvious distinction of the game's magic circle, the clear border between game and reality. This does not mean that players believe themselves to be experiencing reality when playing the game, but instead that they feel empowered in their understanding of history in new, interactive ways and will often seek out new opportunities to learn more about the time period. Summers articulates that historical gameplay requires that "the player adopts different playing styles to match the aesthetic tone of the game: music guides the 'playing posture' of the gamer—how they should play, and how players believe the game will react to their actions,"[45] essential in action/adventure games. In the case of *Odyssey* and *Hades*, the signifiers of Ancient Greekness establish the game's time and place by suggesting to the player that their gaming expectations will conform to their historical knowledge (be it accurate or not). Historical accuracy is not the point. This referent instead provides a framework for understanding and contextualizing expected in-game behaviours as well as supporting effective and immersive gameplay.

CREATING NARRATIVE INFLECTION

Assassin's Creed: Odyssey uses sound not only to signal its Greekness, but also to create narrative inflection. The first cut scene of the game introduces the historical setting of the game, with King Leonidas of Sparta shown looking out over a clifftop before the Battle of Thermopylae in 480 BCE (figure 1.8).[46] Speaking with one of his soldiers, his dialogue belies the battle just to come. Set in front of a backdrop of mountainous terrain, Leonidas speaks of fishing with his son while standing in the rain.

Figure 1.8. Screenshot from the opening cut scene, *Assassin's Creed: Odyssey*. **Source:** Reproduced by permission of Ubisoft Entertainment SA.

Despite the beauty of the terrain and the innocence of his words, however, the imminence of the battle is clear, not because of the visuals or the dialogue, but because of the musical subtext. The melody repeats the same three-note horn motive (B♭, B♭ an octave higher, and A) several times, a motive that does not resolve back to the tonic of the key (D, established through repetition later in the passage). This repetition and lack of resolution increases the tension upon each iteration of the motive, reflecting the emotional tension in the narrative.

This is not the only feature of the soundscape that raises the intensity of the scene towards the battle to come. Figure 1.9 transcribes several elements of the scene's soundscape, including volume, tonality, instrumentation, and dialogue, in conjunction with the visual and narrative events presented on-screen. At the fourth repetition of the horn motive, Leonidas spaces out his dialogue and musical elements begin to increase in density. At 0:28, for example, a quiet cymbal swell is heard, leading into a bass drum pulse that begins at 0:32 and continues growing in volume (as shown by the crescendo-like shape of this event on the figure) through the introduction of other instruments and musical motives. Low brass and string instruments are introduced at 0:42 to fill out the lower register of sound, and alternate between pitches D and C. At the same time, the music shifts away from a solitary drone tone to introduce dissonances, with a particularly prominent major second interval heard between D and C over other background tones in the low strings.[47] Finally, when Leonidas puts on his helmet and the camera pans to the harbour, showing a fleet of attacking ships, the music swells in volume, arriving at its loudest point by 0:48. The music emphasizes the dominant scale-degree in the bass, and the bass drum continues to grow louder while articulating a steady percussion pulse at the rate of a fast heartbeat. The rise in density, volume, and dissonance are all intended to create a sense of tension and anticipation in the player for the battle to come, using modern musical idioms rather than historical ones to accomplish this.

Changes of viewpoint and scene continue to be supported through the soundscape as the cut scene continues. At 1:16, for example, the volume drops in synchronization with a shift in camera angle, followed by a quick ramp-up in sound. The instrumental density once again increases, with a new dissonant semitone motive introduced in the high violins. This time, the character dialogue also supports the increase in intensity, a more literal representation of narrative inflection. Leonidas's voice is pitched much higher in his range, and by 1:30 he participates in a call-and-response dialogue with his troops punctuated by the soldiers' grunts and weapons hitting the ground. The scene culminates in intensity at 1:43 with three instrumental groups layered on top of one another where the enemy charges. At 1:50, all visuals and audio fade out to cue the player that the scene is ending and a new one is about to begin.

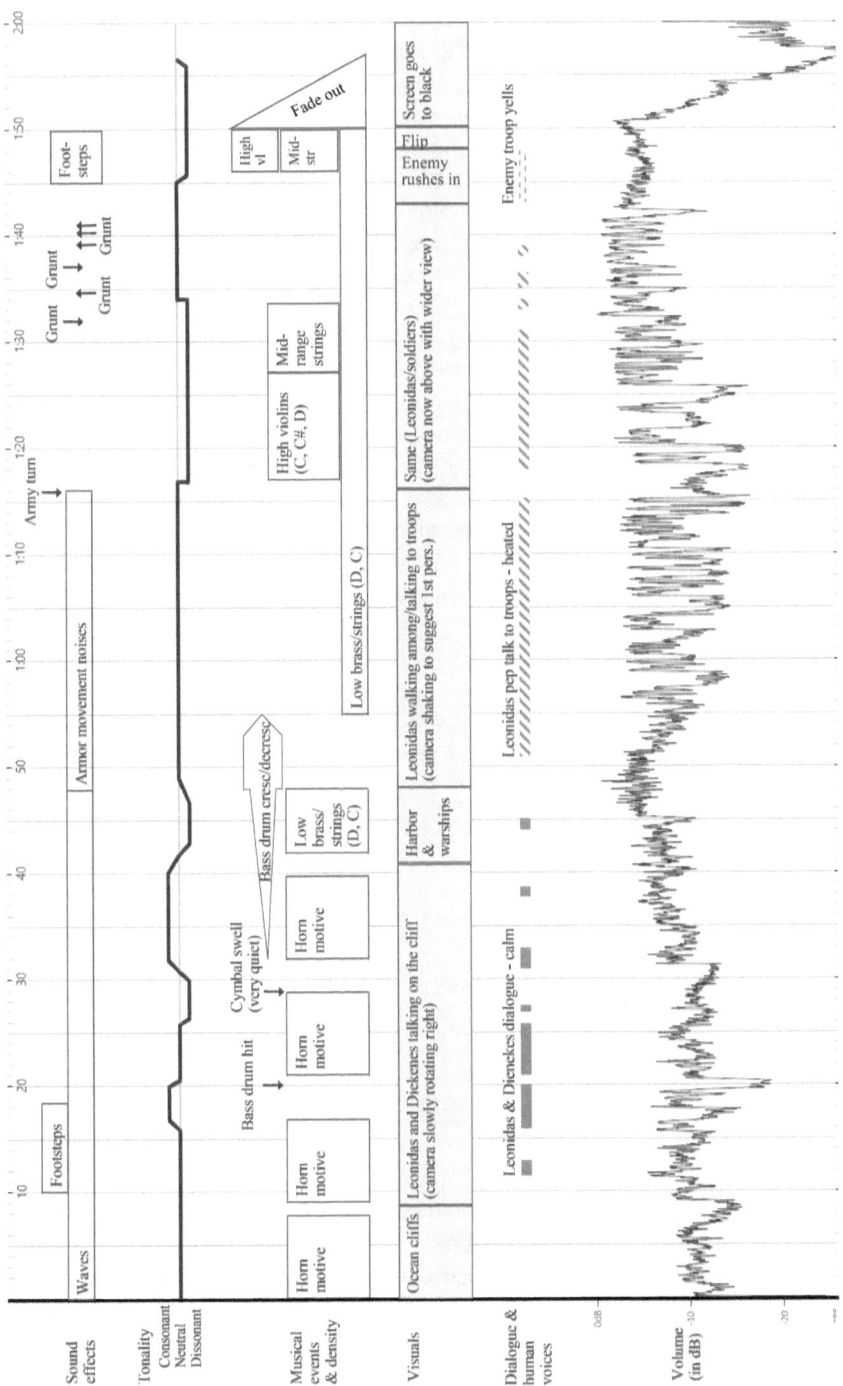

Figure 1.9. Analysis, *Assassin's Creed: Odyssey*, Leonidas opening cut scene. *Source:* Figure by the author.

The scene highlights a tension between the (diegetic) sounds of the gameworld itself—the dialogue, the sound of the waves, soldiers' footsteps and weapons—and the (non-diegetic) musical cues. The result is a musical subtext that articulates the anxiety of the characters in preparation for the battle to come, thereby generating tension with the dialogue and providing nuance to the overall narrative being communicated. While on the surface this may not seem like gameworld at all—in contrast to our other scenes from *Assassin's Creed: Odyssey*, this example reinforces little to no musical features articulating Ancient Greece—in many ways the sounds *do* establish the characters and their status as believable people within the world. The reliance on heavy strings and brass, coupled with a rapid rhythmic pulse, is the epic style oft-referenced by players and scholars. Summers in particular argues that these act to "texture" the gameworld to give it more depth and believability, and that the concept of epic is often associated with games that draw inspiration from classical mythology but is not limited to that context.[48] Interestingly, Summers cites the heartbeat effect as a common means of creating epic structure, and states that this heartbeat in music "connects the first-person incarnation of the avatar's body to the character's role and position within the wider context."[49] While he does not use the term epic to identify specific musical properties, he does identify:

> The standard battle cue is a driving *allegro-vivace* piece that uses a synthesized orchestra in a heroic-military topic based primarily on brass *marcato* interjections and fanfare gestures (accompanied by snare drum, cymbal, tambourine, and other percussion). The musical cue stimulates my psychological arousal—I am primed to react and I am aware that it is own to my personal skill whether the heroes' party will defeat the enemy.[50]

This suggests that the music evokes a gameplay function in addition to its role in creating narrative inflection, and in turn suggests that such a function might be applied more broadly to games in a wide array of other genres.

GAMEWORLD CUEING

In the survival horror genre, sound is particularly used to establish space, but in a different context than the historical games described above. In addition to establishing geographical and temporal elements, sound is used to create a both virtual sense of proprioception as well as a "sixth sense" of safety versus danger. Horror has been a particular interest of many game music scholars, with notable works by William Cheng, Alexander Kolassa, Rebecca Roberts, Guillaume Roux-Girard, Isabella van Elferen, and Zach

Whalen outlining the psychological and musical impacts of the genre.[51] A particular focus has been the popular game series *Silent Hill*. Florian Mundhenke and Zach Whalen, for example, both discuss the various roles of sound and music within the *Silent Hill* series and how film and game adaptations differ in their treatment of soundscapes due to varying functional priorities.[52] William Cheng, in chapter 3 of *Sound Play*, goes in depth into *Silent Hill*'s unpredictability, how this manifests in soundscape, and how this differs significantly from sound and music in other game genres. Despite discussion in these sources of how *Silent Hill*'s soundscape differs significantly from that of other games, the game's treatment of sound has become a model for many others within the horror genre, and as such many of the observations made by Cheng, Whalen, and Mundhenke in particular will elucidate the role of sound in horror in this chapter. Sound and music do not exist simply to create the mood (although that is a significant function), but rather, interacting with sound is an essential component to effective gameplay. As Rebecca Roberts states, "survival horror games cannot operate to their full potential without sound and music."[53]

In this section, I will analyze a gameplay scene from *Metro: Exodus*, a first-person shooter with survival horror elements released in 2019.[54] After an opening cut scene that establishes backstory for the game's post-apocalyptic world, the player begins their interactive gameplay underground in a network of tunnels that form the game's main play area. The sounds in this portion of the game are primarily diegetic, with creaking doors, water dripping, and footsteps acting to establish time and place. The lack of music and the focus on diegetic rather than non-diegetic sound found in *Exodus* are common features of survival horror games. As Mundhenke explains in his own analysis of *Silent Hill*, "since fear and confrontation is already set up by the noise: music, as stated above, would possibly have overdone the scene and could have had a distracting effect."[55] These environmental sounds set the mood since, as Whalen describes for another game, "actual silence is rare in the game, and the constant blanket of noise, which varies dramatically in volume and complexity, drives the anxiety of the game experience."[56] However, despite the sounds' role in establishing the barren, metallic gameworld and boosting narrative tension, they also fulfill a cueing function within the game.

Figure 1.10 gives an analysis of the first two and half minutes of interactive gameplay from *Exodus*, breaking down various components of the soundscape. Player-generated sounds form much of the sonic environment. Two elements not indicated on the chart because of their constant presence—footsteps and shallow breathing noises—act to inflect the narrative by providing sonic signals of the player's constant state of tension and their apparent aloneness in the underground (soon to be disrupted).[57] However, other player-generated sounds indicated on the chart, such as

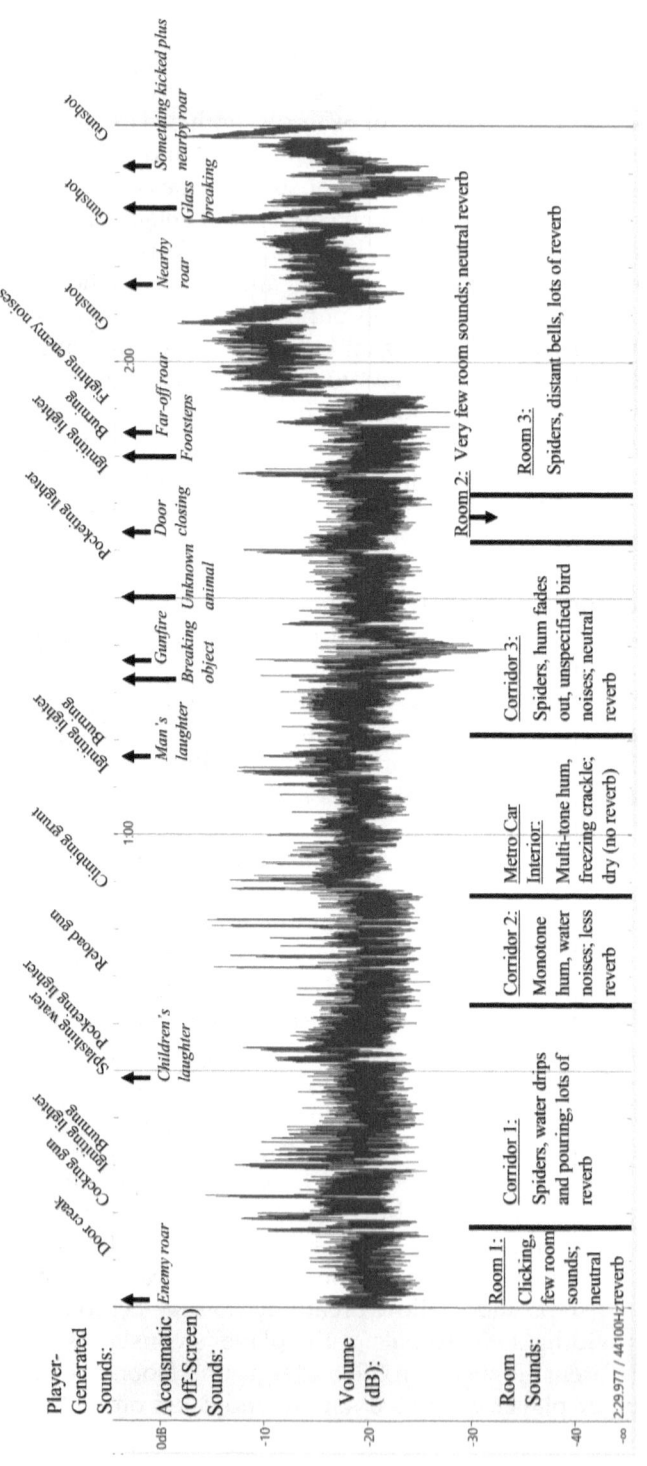

Figure 1.10. Soundscape analysis, opening sequence, *Metro: Exodus*. Source: Figure by the author.

gun cocking, reloading, and igniting or pocketing their lighter, in addition to creating a more textured, realistic gameworld, act as gameplay cues to indicate that the action the player has initiated (through buttons or other controls) has been successful. Player-generated sounds can also give the player cues about their location within the virtual space. At 0:29, for example, the louder sound of splashing water tells the player that they have stepped into a puddle at the base of the far wall caused by a leaking pipe (visible before the sound is triggered), and that they can therefore progress no farther in that direction. This is superimposed upon other water noises within corridor 1, such as drips, that act to texture the environment. As seen by the room sounds indicated on the figure, the combination of sounds is another means through which players can identify (in very dark lighting) when they have entered or exited a specific room or hallway within the level. Julianne Grasso examines a similar musical cueing from a temporal standpoint, stating that "music exists through time, and time in games is organized by narrative events facilitated by play," thus serving a functional role to delineate game state.[58] I would like to suggest a similar delineation here, linking not only to time but also to the gameworld's physical layout: the player progresses through the space both temporally *and* spatially, and different zones are articulated through perceivable differences in the soundscape. And as Grasso argues:

> A player's recollections and recognitions of styles and references, including recognizing the cue itself, informs play and the listening process more broadly. In other words, players need not be paying much attention to the musical processes . . . when these cues become familiar enough to signal all of the meanings and associations that are set to unfold amongst their musical processes.[59]

Upon repeated exposure to the game and its soundscape, the process of recognition becomes automatic and players are able to use these cues to inform game choices such as speed of movement, whether to reload ammunition, or choosing their next direction based on their perception of threat versus safety at their current location. Cueing is thus directly linked to player action.

Similarly, cueing is not always a result of player actions but can act as foreshadowing of future events. The acousmatic sounds in this scene (that is, sounds that suggest an off-screen presence)[60] fulfill all three roles of gameworld sound simultaneously. Sounds such as off-screen children's laughter and the screams of unknown animals, common horror film clichés as identified by van Elferen, suggest a disembodied "uncanny."[61] Furthermore, they are signs to the player that this is a horror game and they can expect many of the conventions of that genre, including unexpected enemies jumping out from dark corners. While Cheng argues that the use of industrial noises combined with howling monsters and worse might "signal a collective disavowal of aesthetic legibility—a disavowal born of trauma so unspeakable that its symptoms and aftershocks lie beyond all conventional representation,"[62] I believe the opposite is true here. Such noises serve to more thoroughly ground the gameworld in our real-life world in order to heighten the sense of the uncanny—of the world not being quite right—when fantastical elements such as demons appear. Indeed, later comments by Cheng support this interpretation. He explains that since "the noises in there [the soundscape of *Silent Hill*] are dead ringers for the everyday noises out here, we are working with scrambled sensoria."[63] Sounds in *Exodus* such as non-player footsteps, a man's maniacal laughter, and far-off roars raise player tension by highlighting this discrepancy from the real world, by suggesting the proximity of future enemies, and also by suggesting possible destinations that the player might take to progress within the game. In fact, in many cases these sounds will increase in volume if the player heads in the correct direction, providing a cue to the player that they are headed towards an area of activity—a concept Mundhenke terms a *navigational function*.[64]

One of the most common methods of cueing via volume in shooters and horror games can be heard starting at 0:46. At this moment an enemy roar is heard, which increases in volume at 2:10 and 2:25. The lower volume cues the player that an enemy will appear shortly, while the louder volume announces the immediate threat. The moments of gunshots by the player that appear on the figure are in response to this, taking down each enemy. The change in volume therefore acts as a sonic signifier, as described by Roberts, a form of gameplay cue that warns the player to get ready for action because a threat will appear imminently.[65] Such distance-indicating sounds are common in horror; for example, a similar cueing function in *Silent Hill* that uses variances in the sound of radio static is cited by Mundhenke, Whalen, and Cheng.[66]

The volume band on the figure shows some notable things about the soundscape of this scene, though. Most noticeable is the spike in volume around 1:54, corresponding to an enemy attack. However, also visually apparent on the analysis is the difference in sonic profile (including volume) within each room. The reverb of the metro tunnels changes as the player-character moves from a flooded area (corridor 1) to one with more metallic surfaces (corridor 2), for instance. In a particularly noticeable contrast, the sound becomes less reverberant and more dry upon entering a colder area (metro car interior), matching visual cues of frost in the environment's metro cars and windows as well as on the player-character's field of view to suggest a freezing helmet. The relatively static volume of corridors 1 and 2 (volume spikes shown on the chart in these areas are the result of player-generated sounds) becomes much more irregular in the metro car interior, and the dip then rise of volume in corridor 3 ramps up to the upcoming enemy encounter in room 3. Although decisions by the player about how long to remain in each room will impact the exact timing and sequence of sounds, as Mundhenke describes, "uncomplicated and non-ambiguous sounds and with relatively static acoustic spaces [are] bound to locations."[67]

Upon further examination, it also appears that the game designers are structuring a particular way of engaging with sound. Environmental sounds alone are heard upon first entering a room, and acousmatic sounds suggesting areas off-screen are held back until the player moves farther into the room. For example, in the gameplay analyzed on the figure, in corridor 1 the children's laughter does not occur until the player has spent ten seconds in the room, and in corridor 3 the acousmatic sounds do not occur for approximately five seconds. The exact amount of time the acousmatic sounds are held back vary depending on player action—how quickly they move into the room—but this structure nevertheless establishes a particular process that allows the player to attune to the new environment first, then progress into larger-scale gameplay observations. The acousmatic and room sounds are much less random than they appear and serve a greater gameplay function of directing the player onwards after they have processed their new environment. This may have an impact on the perception of acousmatic sound, supporting Collins's observation that in-game movement can "de-acoustamize" sound as players approach a sound's source, effectively allowing them to engage with and modify the sound.[68]

CONCLUSION

Much of this discussion about gameworld has focused on realism in its various manifestations. Do anachronisms or a lack of realism disrupt this relationship between the game and its players? Van Elferen explains that the role of the game soundtrack is to draw the player into the atmosphere of the game space,[69] but that does not necessarily mean that accuracy is essential. In fact, authenticity, as understood by *players* in these two genres, does not describe whether the gameworld matches our real-life expectations of time and place but instead more commonly indicates how thoroughly the player "buys in" to the fiction presented to them. Historically accurate music (objective authenticity) may act more as a disruption to the player experience if the style and genre are completely foreign to them. Sounds provide a frame of narrative and gameplay expectations, as described by Whalen and Taylor:

Video games provide unique opportunities for conveying ideas because the environment a game constructs allows the designer to build in a set of relationships (an ideology, for example) even before the player confronts a narrative based in that environment. The effects of these changes on historical representation operate similarly to nostalgia's desired return to a particular moment or to a mythical state of innocence. In this context, play becomes a way of relating to the past and a means of addressing our loss of innocence by temporarily allowing us access to that innocent condition.[70]

As we have seen, the three functions for gameworld sound presented here, while articulated as three distinct components, more commonly engage with and build off one another to create a rich, emotional, engaging world. In the games presented within this chapter, players conceptualize space in a way that is intimately integrated with their listening, supporting Collins's argument that sound plays a fundamental role in our imaginary construction of virtual space.[71] But worldbuilding (by establishing time and place), texturing characters and storyline (through narrative inflection), and event cueing occur simultaneously and are deeply entwined both at a functional level and in terms of players' perceptions of immersion (being in the game) and successful gameplay realization. Objective and constructive authenticity thus exist simultaneously, intermingling to create the overall player experience.

NOTES

1. Zach Whalen and Laurie N. Taylor, *Playing the Past: History and Nostalgia in Video Games* (Nashville, TN: Vanderbilt University Press, 2008), 2.

2. An earlier form of the Greek music analysis of this chapter was presented at the 2021 North American Conference on Video Game Music. Many thanks to everyone who gave feedback and suggested additional resources (particularly those on Greek music) to investigate, but especially Gregg Rossetti and Dan Donnelly.

3. Katie Salen and Eric Zimmerman, *Rules of Play: Game Design Fundamentals* (Cambridge, MA: MIT Press, 2003), 401–402.

4. Tison Pugh and Angela Jane Weisl, *Medievalisms: Making the Past in the Present* (London: Routledge, 2013), 127.

5. Van Elferen, "¡Un Forastero!," 36.

6. Ernest Adams, *Fundamentals of Game Design*, 2nd ed. (New Riders, 2010), 84.

7. Collins, *Playing with Sound*, 45.

8. van Elferen, "¡Un Forastero!," 30.

9. William Cheng, *Sound Play: Video Games and the Musical Imagination* (New York: Oxford University Press, 2013), 99.

10. See, for example, Zach Whalen's citation of film music function from Aaron Copland in "Case Study: Film Music vs. Video-Game Music: The Case of *Silent Hill*," in *Music, Sound and Multimedia: From the Live to the Virtual*, ed. Jamie Sexton

(Edinburgh: Edinburgh University Press, 2007), 70; and David Neumeyer and James Buhler's referential, expressive, and motivic markers of sound (which they imply are narrative functions in that text, but will be adapted here as both narrative and gameplay functions) in *Meaning and Interpretation of Music in Cinema* (Bloomington: Indiana University Press, 2015), 11.

11. Aris Politopoulos, Angus A. A. Mol, Krijn H. J. Boom, and Csilla E. Ariese, "History Is Our Playground: Action and Authenticity in *Assassin's Creed: Odyssey*," *Advances in Archaeological Practice* 7, no. 3 (2019): 317–319.

12. See, for example, https://www.vice.com/en/article/59kwea/a-harvard-egyptologist-explain-the-historical-accuracy-of-assassins-creed-origins and https://www.pcgamesn.com/assassins-creed-odyssey/historically-accurate.

13. *Assassin's Creed: Odyssey* (Ubisoft Entertainment, 2018), Microsoft Windows, PlayStation 4, Xbox One, Nintendo Switch, Stadia, music by The Flight, Mike Georgiades, and Giannis Georgantelis.

14. Gibbons, *Unlimited Replays*, 31.

15. Politopoulous et al., "History Is Our Playground," 321.

16. Ubisoft, "Composing an Epic Score for *Assassin's Creed Odyssey*," *news.ubisoft.com* (October 5, 2018), https://news.ubisoft.com/en-us/article/5Y0vBS44xVo7HJocYvb5Ej/composing-an-epic-score-for-assassins-creed-odyssey, accessed July 22, 2020.

17. Hektor Apostolopoulos, "The Flight Share Thoughts on Their *Assassin's Creed Odyssey* Soundtrack," *Viralbpm* (July 20, 2018), https://viralbpm.com/2018/10/07/the-flight-share-thoughts-on-their-assassins-creed-odyssey-soundtrack/, accessed June 12, 2021.

18. "*Assassin's Creed Odyssey* (Original Game Soundtrack) | The Flight," accessed February 18, 2021, https://www.youtube.com/watch?v=fwthw9Sy_RU&t=72s.

19. Stathis Gauntlett, "Antiquity at the Musical Margins: Rebetika, 'Ancient' and Modern," *Byzantine and Modern Greek Studies* 39, no. 1 (March 1, 2015): 110–111.

20. See, for example, Ben Kuchera, "It's Time to Take the Animus Out of *Assassin's Creed*: It's the Vegetables of the *Assassin's Creed* Universe," *Polygon*, last updated October 3, 2018, https://www.polygon.com/2018/10/2/17926100/assassins-creed-odyssey-animus.

21. At the time of publication, this track was available from Ubisoft at https://youtu.be/fwthw9Sy_RU?t=868.

22. Epic is defined in many ways in film and video game music; for the purposes of this study, my understanding of the epic musical trope will be derived from Frank Lehman's outline of the *marcato motto* epic action sequence in Hans Zimmer's film scoring: "significant, portentous, heroic, incredible. . . . A short, loud, and catchy melodic phrase, scored for either strings or full tutti, with an emphasis on brass, though, interestingly, rarely trumpets. . . . Successive notes in these motifs are articulated heavily and separated cleanly from their neighbours . . . louder, more heavily produced, and more densely orchestrated." In Frank Lehman, "Manufacturing the Epic Score: Hans Zimmer and the Sounds of Significance," in *Music in Epic Film: Listening to Spectacle*, ed. Stephen C. Meyer (New York: Routledge, 2017), 31, 36–37. For a deeper perspective on the application of

the epic style to video game, see chapter 3 in Tim Summers, *Understanding Video Game Music* (Cambridge: Cambridge University Press, 2016).

23. At the time of publication, a recording of the sea shanties mentioned in this chapter were available from Ubisoft via Spotify at https://open.spotify.com/album/5lqtcvN17BRAKLEoK6aKbw.

24. One of the major points of note regarding modes is that their names have shifted over time to refer to different scales between the Ancient Greek period, the medieval period, and today. In this chapter, the modern mode names will be used unless otherwise specified. Terms referring to the main orienting note of a mode have also changed over time. While modern-day usage might colloquially use the term *tonic* due to its familiarity among musicians fluent in major/minor tonality, and medieval practice used the term *final*, M. L. West cites further difficulty in teasing out differences between mode, ambitus, and tuning in the historical record of Ancient Greek music. For these reasons, I will use the modern term *pitch centre* or simply refer to scale degree to refer to this concept of tonic or final, and *focal pitch* to refer to pitches that achieve emphasis through repetition or other form of accentuation (which may or may not correspond to the pitch centre of the mode). M. L. West, *Ancient Greek Music* (Oxford: Clarendon Press, 1992), 160–189.

25. West, *Ancient Greek Music*, 131.

26. Ibid., 136–37. A significant amount of West's study catalogues meter in terms of its relationship to text and poetic verse; see pages 136–53 for his in-depth discussion of this topic.

27. West, *Ancient Greek Music*, 206.

28. West, *Ancient Greek Music*, 164. I also wish to acknowledge here that I have intentionally chosen not to reference Stefan Hagel, *Ancient Greek Music: A New Technical History* (Cambridge: Cambridge University Press, 2009), in this discussion due to the arguments presented by Alison Laywine in regards to his interpretation of mode and modulation. For more information, see Alison Laywine, "Ancient Greek Music: A Technical History by Stefan Hagel," *Aestimatio* 9 (2012): 124–170.

29. Nikos Ordoulidis, "The Greek Popular Modes," *British Postgraduate Musicology* 11 (2011): 4.

30. Ibid., 7.

31. See, for example, the various scales shown at https://bouzoukispace.com/category/theory/laikoi-dromoi-book/.

32. Apostolopoulos, "The Flight."

33. Or, as Harry Partch explains, "it is quite impossible for us Westerners to imagine what ancient Greek music was really like, even after we know the salient facts about it" (7). His own study of Ancient Greek music is a notable work of scholarship building an overall study of tuning systems and a compositional method grounded in historical courses from Ancient Greece, China, the Arabic world, and beyond. Partch also cites the importance of speech intonation in several cultures, including the Ancient Greeks; his understanding of this music as "corporeal" emphasizes its connections to both text, drama, and dance. See Harry Partch, *Genesis of a Music: An Account of a Creative Work, Its Roots and Its Fulfillments*, 2nd ed. (New York: Da Capo Press, 1974). Thanks to Gregg Rossetti for bringing this source to my attention.

34. Summers, *Understanding Video Game Music*, 110.
35. Ibid., 115.
36. Ibid., 97.
37. *Hades* (Supergiant Games, 2020), MacOS, Microsoft Windows, Nintendo Switch, PlayStation 4, PlayStation 5, Xbox One, and Xbox Series X/S, music by Darren Korb.
38. The full soundtrack for the game is available from the publisher on YouTube at https://youtu.be/3GRKJ87S5cI.
39. Darren Korb, "We Are Supergiant Games, Creators of Hades, Pyre, Transistor, and Bastion. AMA!," Reddit, September 22, 2020, https://www.reddit.com/r/NintendoSwitch/comments/ixri6b/we_are_supergiant_games_creators_of_hades_pyre/g68cdou/.
40. Additionally, the bass, while more audibly articulating 4/4 due to its strong emphasis of the downbeat via the anacrusis and downbeat arrival, also plays a 3+3+2 rhythm in the first three notes of each bar (albeit diminuted).
41. West, *Ancient Greek Music*, 67.
42. Korb, "We Are Supergiant Games.
43. Gibbons, *Unlimited Replays*, 23.
44. Summers, *Understanding Video Game Music*, 93.
45. Ibid., 110.
46. The scene is available to view on YouTube at https://www.youtube.com/watch?v=6-XWMcsUzLk.
47. On the figure, the "tonality" category is assessed on three levels: neutral indicates that a drone tone or tonic is heard, consonant indicates that consonant intervals (m3, M3, P4, P5, m6, M6) are heard most prominently and/or that perceivable chords within the key occur, and dissonant indicates that dissonant intervals (m2, M2, A4, d5, m7, M7) are heard most prominently. This differentiation is inspired by Elizabeth Medina-Gray's analytical approach and rubrics in "Analyzing Modular Smoothness in Video Game Music," *Music Theory Online* 25, no. 3 (October 2019).
48. Summers, *Understanding Video Game Music*, 64.
49. Ibid., 68.
50. Ibid., 166. This also matches Lehman's definition of the *marcato motto* topic as previously outlined.
51. William Cheng, "Monstrous Noise: *Silent Hill* and the Aesthetic Economies of Fear," in *The Oxford Handbook of Sound and Image in Digital Media*, ed. Carol Vernallis, Amy Herzog, and John Richardson (Oxford: Oxford University Press, 2013): 173–190; Alexander Kolassa "Hail the Nightmare: Music, Sound and Materiality in *Bloodborne*," *The Soundtrack* 11, no. 1 (2020): 23–38; Rebecca Roberts, "Fear of the Unknown: Music and Sound Design in Psychological Horror Games," in *Music in Video Games: Studying Play*, eds. K. J. Donnelly, William Gibbons, and Neil Lerner (New York: Routledge, 2014), 138–150; Guillaume Roux-Girard, "Listening to Fear: A Study of Sound in Horror Computer Games," in *Game Sound Technology and Player Interaction: Concepts and Developments*, ed. Mark Grimshaw (Hershey, PA: Information Science Reference, 2011), 192–212; Isabella van Elferen, "Sonic Descents: Musical Dark Play in Survival and Psychological Horror," in *The Dark Side of Game Play: Controversial Issues in Playful Environments*, eds. Torill

Elvira Mortensen, Jonas Linderoth, and Ashley ML Brown (New York: Routledge, 2015), 226–241; and previously cited Whalen, "The Case of *Silent Hill.*"

52. Florian Mundhenke, "Musical Transformations from Game to Film in *Silent Hill,*" in *Music and Game: Perspectives on a Popular Alliance,* ed. Peter Moorman (Wiesbaden: Springer Fachmedien Wiesbaden, 2013), 107–24; and previously cited Whalen, "The Case of *Silent Hill.*"

53. Roberts, "Fear of the Unknown," 141.

54. *Metro: Exodus* (Deep Silver, 2019), Microsoft Windows, PlayStation 4, Xbox One, Stadia, Luna, PlayStation 5, Xbox Series X/S, Linux, MacOS, music by Alexei Omelchuk. The scene analyzed here has been taken from the following game recording: https://youtu.be/qxLpPVP4m5A. An official live-stream is also available from 4A Games at https://youtu.be/nQlW5U4syxE?t=996 (our scene begins at 16:36 of the official live-stream).

55. Mundhenke, "*Silent Hill,*" 115. Roberts, "Fear of the Unknown," also discusses several examples of musicless horror gameplay in her study, although on the whole she acknowledges the fundamental role of music in horror.

56. Whalen, "The Case of *Silent Hill,*" 78.

57. Roberts cites a similar footstep effect in her analysis of the game *Limbo*: "Certain sounds take prominence within difference sequences, most notably in the presence of loud footsteps, which work to underscore the sheer silence and emptiness of the environment" ("Fear of the Unknown," 149).

58. Julianne Grasso, "Music in the Time of Video Games: Spelunking *Final Fantasy IV,*" in *Music in the Role-Playing Game* (New York: Routledge, 2019), 98.

59. Grasso, "Music in the Time of Video Games," 108–109.

60. As defined by Michel Chion in "The Acousmêtre," in *Critical Visions in Film Theory: Classic and Contemporary Readings,* eds. Patricia White, Meta Mazaj, and Timothy Corrigan, 156–165 (Boston: Bedford/St. Martin's, 2011); and van Elferen, "Analysing Game Musical Immersion," 41.

61. van Elferen, "Analysing Game Musical Immersion," 41; and "Introduction: Sonic Horror," *Horror Studies* 7, no. 2 (2016): 166–167.

62. Cheng, *Sound Play,* 95.

63. Cheng, *Sound Play,* 102.

64. Mundhenke, "Musical Transformations," 115.

65. Roberts, "Fear of the Unknown," 145.

66. Cheng, *Sound Play,* 109; Mundhenke, "Musical Transformations," 115; Whalen, "The Case of Silent Hill," 76.

67. Mundhenke, "Musical Transformations," 114.

68. Collins, *Playing with Sound,* 49.

69. van Elferen, "!Un Forastero!," 30.

70. Whalen and Taylor, *Playing the Past,* 12.

71. Collins, *Playing with Sound,* 45, 54.

Two

Tropes

In *Playing the Past: History and Nostalgia in Video Games,* Zach Whalen and Laurie Taylor discuss the effect of previous game and media experience not only on the player but also on game designers. Designers, according to the two authors, use *schemas*—that is, patterns of expectation established through previous gameplay experiences—to both guide and limit player behaviours. In Whalen and Taylor's interpretation, the degree of similarity between the gameworld and the real world has a direct impact, with worlds similar to our own prompting "far greater expectations of freedom" from players because of the open-ended nature of the real world.[1] From an authenticity standpoint (specifically, that of constructive authenticity, which includes player expectation and perception), it appears that players thus expect a repeated crossing of the magic circle's boundary, the mental delineation of the game space. There is a clear concept of what constitutes the game, its rules, and the parameters of the gameworld, but this is built upon a scaffold clearly founded on players' knowledge of the world outside the game. Players consequently repeatedly refer to real-life experiences to structure their virtual play, a crossing of the magic circle boundary that merges the real and the imaginary.

This cross-boundary movement reinforces flow and a sense of agency, but these are not the only advantages of evoking the real world, especially when we discuss the perspective of game designers rather than players. William Gibbons, as we discussed last chapter, cites such references as a form of cultural shorthand that allows us to reference real-world locations.[2] Quite simply, this process gives game designers a shortcut for designing their gameworlds since, rather than building the entire fiction from the ground up, references to previous worlds allow them to quickly suggest particular environments to the player. And such processes are not limited to gameworld. Tropes are one method through which game designers can provide a quick shorthand to stock narrative conventions, game environments, genre, and more. While this subject has been examined much in scholarship on the music of both film and video games, the definitions and vocabulary used lack standardization. Whalen and Taylor, for example, may use *schema* to describe a very similar concept to what Robert Hatten, David Neumeyer, and James Buhler term *trope*.[3] Isabella van Elferen describes media literacy—that is, an audience's knowledge from previous media and how that impacts their understanding of new media—in a way that evokes schema, trope, and intertextuality.[4] I would like to suggest here that all of these terms are nuanced reflections of the same concept: Modern-day video game culture is strongly linked to a wider media culture formed by our shared experiences witnessing film, television, game, and more recently memes and social media. While players will inherently have individualized intertextual experiences given their widely disparate lived experiences, one major facet of this twenty-first-century global media culture is the act of playing on expectations established through previous media consumption, whether that involves adhering to expected plot devices or subverting them. Player agency and thus their perception of authenticity (in this case, constructive authenticity) is invoked by giving them the power to recognize tropes and to engage with them.

This chapter will explore this cultural phenomenon of trope, building on previous work by both film music and game music scholars. Scholarship in video game music has been keenly interested in tropes, with notable writings by Sean Atkinson, Karen Cook, and Tim Summers and numerous conference presentations, particularly at the North American Conference on Video Game Music.[5] This chapter will focus on musical instantiations of trope, how our implicit knowledge of music theory impacts our perception of these, and how other game factors such as visuals and narrative intersect with the interpretation of tropes. Music theory is an area that is commonly thought to be the domain of experts, but I would argue that even those without formal music knowledge subconsciously pick up on traits often associated with music theory, such as divisions of musical form, elements of contrast, and consonance/dissonance in melody and harmony. Players might describe these concepts emotionally rather than concretely—associating concepts such as rhythmic irregularity or dissonance with tension and anxiety, for example—but embedded within these emotional observations are expectations of specific structural patterns in music. The analyses in this chapter will begin with a few short, well-known examples from video games and then continue with a case study of the Fantasy genre and how it plays on player expectation. The chapter will wrap up with a discussion of some problematic aspects of trope implementation. Before we get to the analysis, through, let us clarify some definitions, which are muddy at best in the current literature.

TROPES: NARROWING DOWN A DEFINITION

What exactly is a trope? Although the term has long-standing usage in the study of music dating back to the medieval era, our current understanding of the term has more in common with studies in rhetoric and semiotics with their common interests in the study of language and meaning.[6] In the study of rhetoric, tropes are understood as figures of speech that function as basic structures by which we make sense of experience.[7] We use many such tropes in writing and in spoken language (metaphor and irony, for example), and they feature an abstract form of representation rather than a direct description of the object in question. The study of semiotics builds on this definition but takes it one step further. Semiotics explores this connection between object and representation in its quest to understand how meaning is created, understanding an object (a *sign*) to have an associated meaning and certain features that represent it (*signifiers*). Tropes in these two fields of study therefore indicate a relationship between a sign and its mental concept, between signifier and signified.

The study of tropes in music scholarship has taken a different path, with its roots in the development of classical topic theory in the 1980s. As Danuta Mirka describes, Leonard Ratner's seminal 1980 work on style and affect *Classic Music: Expression, Form, and Style,* led to further developments by Kofi Agawu, Robert Hatten, and Raymond Monelle rooting topic theory in semiotics. Topic theory, however, has mainly had a historical focus, with much of this work centered on the music of the eighteenth century. As Mirka describes, "the semiotic status of topics has been framed in terms of modern semiotics developed by twentieth-century authors such as Charles Sanders Peirce, Ferdinand de Saussure, Roman Jakobson, and Umberto Eco, but the eighteenth century possessed its own theory of signs that imbued music aesthetics."[8] Hatten, unlike many of the other topic theory scholars, has shifted the focus from *topic* to *trope* due to his interest in linking to parallels in language and semiotics. As he explains, "an eighteenth-century musical topic is a familiar figure, texture, genre, or style that is imported into a new context, where it may interact with other topics and the prevailing discourse of a movement in ways that may spark creative meanings. These interactions are analogous to the processes that produce tropes in literary language."[9]

However, this semiotic-inspired definition does not quite get at the heart of trope's pop culture definition, the definition that in my view is the most useful for discussing video games. While the two are certainly related, popular culture defines tropes more colloquially as a means of indirectly conveying a concept to the audience, stock ideas that audiences learn from their previous exposure to other media. Perhaps the best-known website cataloguing this phenomenon is TVtropes.org. Calling themselves "the all-devoring pop-culture wiki," TVtropes.org is an archive of sorts, cataloguing recurring references within film, television, books, video games, and other media sources. For those unfamiliar with the website, users may look up entries such as "The All-Loving Hero," "The Blind Alley" chase scene, and so forth to view a member-compiled list of films and other media that employ particular narrative devices, genre conventions, and more. Each trope page includes a brief description of the narrative convention along with a catalogue of references. TVtropes defines the term as follows:

> A trope is a storytelling device or convention, a shortcut for describing situations the storyteller can reasonably assume the audience will recognize. Tropes are the means by which a story is told by anyone who has a story to tell. We collect them, for the fun involved.
>
> Tropes are not the same things as clichés. They may be brand new but seem trite and hackneyed; they may be thousands of years old but seem fresh and new. They are not bad, they are not good; tropes are tools that the creator of a work of art uses to express their ideas to the audience. It's pretty much impossible to create a story without tropes.[10]

This new definition adds something that the definitions from rhetoric and semiotics do not quite grasp: Tropes are tools or building blocks for the creator, and they are not just present but *expected* in modern media. Tropes according to this definition are not new to visually based musical media, but instead have been a longstanding narrative device in opera and film, for example, linking to stock plot devices in literature and folk tales.[11] And furthermore, despite their differences in definition, Neumeyer/Buhler and TVtropes both define tropes as understood *intuitively*, with the viewer not needing to work at figuring out some sort of hidden message. Rather, they are meant to immediately recognize the storytelling convention being invoked.[12] I would also argue that tropes are understood *culturally* since without the framework of a particular canon of media—that is, a particular body of tropes—viewers read the media quite differently. Take, for example, one trope outlined by TVtropes, the use of local languages in Bollywood cinema. To a viewer unfamiliar with Bollywood cinema, the use of local languages would simply seem natural given that the film was created in India for Indian audiences.[13] However, to Indian viewers, the use of local languages might indicate that the character has "street cred," a connection to local community, or that they fit in with a particular social class. This nuance and set of expectations are lost on those not familiar with the language or culture since differences in dialect, tone, and expected social conventions are not perceptible by those audiences.

TVtropes also captures the idea that tropes are fluid and ever changing, stating, "TV Tropes doesn't get to set what the term means; the best we can do is capture the way it is used. Since there's no consensus on a precise definition, the best way to describe the phenomenon is by example."[14] As a result, the perception, realization, and understanding of tropes may vary widely over time. For instance, a player in 1992 might interpret the use of pixel graphics, poor-quality sound, and a narrative focusing on rescuing a princess as simply the standard for games of the era, while the same player in 2021 would instead understand these features in a new game as forming a retro gaming trope (a subject we will expand upon later in chapter 4).

Trope and Topic

In this book, my discussion of trope will blend the popular culture definitions with elements from Hatten's approach to topic theory, an approach that will allow a distinction between topics as fixed elements of the sound, music, visuals, or other aspects of the game that merge together to form tropes. The shift to using tropes to study new multimedia (as opposed to its focus on the common-practice music of the eighteenth century) has been led by David Neumeyer and James Buhler, building on Robert Hatten's interpretation of topic theory.[15] Neumeyer and Buhler share Hatten's understanding of trope and topic as paralleling the distinction in semiotics between signified and signifier, but further expand this to state that topics are stable while tropes are "shifting, creative, or altering."[16] Thus tropes are not merely the combination of topics, but rather how the combination generates a new associated meaning through repeated usage and through its interaction between image and sound, a manifestation of Michel Chion's added value principle.[17] Tropes may communicate information that may not be stated directly by character dialogue, such as time and place, but they may also suggest more nuanced concepts such as emotion, tension, reaction, and more, thereby creating "a kind of parallel emotional/aesthetic universe."[18] We essentially have two (or more) modes of communication occurring simultaneously. In games, as in film, tropes are often communicated through music which plays non-diegetically along with character dialogue. Because of this abstract nature, musical tropes typically function nonverbally (with texted song the obvious exception), making music a useful device for more subtle elements of communication such as emotion. Sean Atkinson considers how this is recontextualized for video game, and observes that "occurrences of tropes in music (especially instances such as video game cues) have an *independent* power to convey narrative. . . . This ability to engender narrative allows for the music to bring a rich counter-text to the entire production."[19] Since player choice is a critical component of video gaming, musical cues and other parameters can have a more direct influence on the player than in film. "Manipulations of expected topical norms could influence a player's decisions, having a real and meaningful impact on the experience of the game."[20]

Neumeyer/Buhler's and Hatten's definitions (and Atkinson's adaptations of these for video games) provide some basic language about understanding how musical elements combine to produce tropes, but a more important component that they include in defining trope (and that is not always clear in the scholarly literature) is that a trope's meaning resides in the interaction between its constituent topics, the interrelation of topics to produce particular genres and styles, other elements such as visual images, and the audience's previous experiences. For example, a trumpet playing in arpeggiated tones might produce a fanfare effect, but if we pair that visually with a scene of a medieval castle, the audience will understand this to be a reference to nobility, not because a trumpet fanfare is inherently noble, but rather because they have viewed scenes from television, film, and video games that used the trumpet fanfare in connection with medieval kings and queens. As John Haines describes, this is one of the most common medievalist tropes, one which had an early history in film: "Almost as soon as medieval cinema began, the trumpet fanfare and the horn call were placed on the screen in a stereotypical way: players swiftly raising their instruments to their mouths and blowing."[21] Without this particular cultural memory, though, the reference is lost to the audience and thus the trope is not truly realized since the effectiveness of the trope itself is based on how well known the represented element is to its audience. Additionally, as Iain Hart explains, the communicative capacity of musical signs (and thus tropes) is just as strongly rooted in the player's personal experience as in the game composition itself. If the player has not encountered the trope before, they will not be able to understand what it is trying to communicate.[22]

For the sake of avoiding confusion, from this point forward in this book I will refer to Hatten/Neumeyer's conceptualization as *trope*, and the pop culture definition established by TVtropes.org as *media trope*. Be aware, however, that in some cases it is impossible to separate these two definitions.

TROPES: SOME MUSICAL EXAMPLES

Quotation as Trope: The Toccata and Fugue in D Minor

Let us take a look at a few examples, starting with perhaps the most notorious use of topic and trope in Western horror media. Figure 2.1 gives

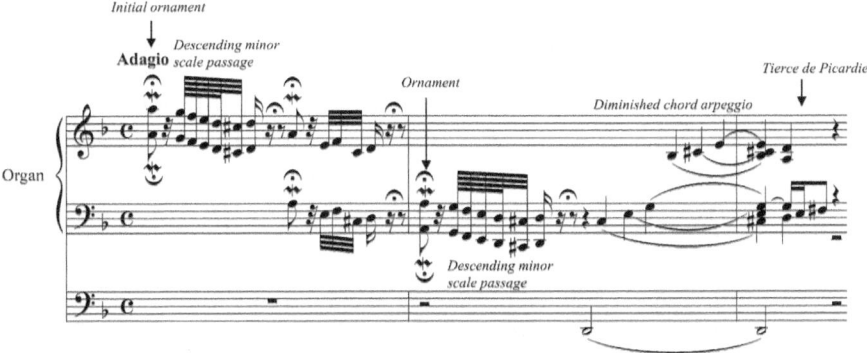

Figure 2.1. The instantly recognizable opening of the Toccata and Fugue in D Minor, BWV 565, by J. S. Bach. *Source*: Public domain. Analysis by the author.

the opening of the *Toccata and Fugue in D Minor*, BWV 565, commonly attributed to J. S. Bach.[23]

First used in connection with the horror genre in 1931 in Rouben Mamoulian's film adaptation of *Dr. Jekyll and Mr. Hyde*,[24] the opening theme of this work has appeared in countless other references to the genre. The song has been a popular choice from the early days of video gaming. Dana Plank, for example, cites at least fourteen games between 1982 and 1993 that use excerpts from the work.[25] Plank rightfully indicates that the opening incipit was most frequently used by game designers, but also that this motivic quotation was often transformed—intentionally or unintentionally—to position the motive in new artistic contexts that nevertheless retained the motive's allusions to gothic horror.[26] Most audiences are completely unfamiliar with its origins as a baroque-era organ work, instead simply associating it with horror film. Interestingly, Gibbons discussed his own childhood experience encountering the motive in *The Battle of Olympus* (1988), describing how his previous familiarity with the tune created a disconnect when encountering it in the context of the new game:

> I distinctly remember my confusion at hearing this piece of music, which led me to set the controller down for a few minutes, both to listen and to think. I knew this piece from Disney's *Fantasia* (1940), and I had even played it myself in piano lessons (in a much simplified version). What on earth was this music doing in a video game? What did Bach, this particular piece, or classical music in general, have to do with Greek gods and heroes [in *The Battle of Olympus* (1988)]? I was sure this music *meant* something—maybe even something profound—but I just couldn't put my finger on exactly what it was.[27]

Gibbons's sense of disconnect with the musical reference might have resulted from the song's use in an atypical genre (Greek mythology rather than horror, a discrepancy compared to previous media associations with the *Toccata*) or perhaps from his centering of the work as part of the canon of music learned in his classical-centered piano lessons. The horror meaning does not exist inherently in the work itself, but rather is strongly connected to how the motive has taken on new life as a (clichéd) horror theme rather than as a baroque-era historical artifact.

For me, the *Toccata*'s actual usage as a trope/media trope *beyond* the opening incipit falls somewhat flat since audiences don't recognize the remainder of the work—or those who do are more familiar with the work from a performance or music history background, a perspective that will color their perception of the motive quite differently. Take its appearance in *Dark Castle* (1986, re-released for Sega Genesis in 1991), for example.[28] The game begins with the recognizable initial ornament, descending minor scale passage, diminished-chord arpeggio, and resolution to the Tierce de Picardie, the elements that might be termed the topics of this motive. Entering over the opening credits, the motive effectively establishes the horror trope not by its inherent musical properties, but rather through its association with the game's visual and narrative setting: An adventurer progresses through a stony, dark castle colored in greys, blacks, and dark blues while killing bats, rats, zombies, a torturer, and other enemies typical of the horror genre.[29]

Additionally, in this instance the music progresses well beyond its opening incipit to less well-known portions of the work. For me, this serves a reminder of learning the work during my teenage years that distracts me from its attempt to invoke Gothic horror references. Given my personal cultural knowledge, this motive does not subtly invoke the horror genre through its previous accumulated meaning, but rather comes across as a decidedly blunt and cheesy way of evoking horror, disrupting my focus on the gameworld.

Theoretical Process as Trope: Shadow of the Colossus

Tropes and topics can, however, be used much more subtly, and modern game composers, with their access to more realistic sounds, a wider array of world instruments, multitrack scoring, and extensive audio after-effects have taken advantage of these resources. And as Plank describes, topics and tropes need not only be established by preexisting musical themes, but instead the use of particular musical timbres, textures, chords, keys, rhythms, and more can allude to the player's previous experience, combining to create new meaning.[30] Let us take a look at an example from 2005's *Shadow of the Colossus* (re-released in 2018).[31]

The game is often cited as notable for its lack of background music, a subject Gibbons explores in depth, citing silence as a key element reinforcing the sense of isolation and moral ambiguity within the game.[32] The music of the game, according to Gibbons, works in juxtaposition with its strong emphasis on silence: "The music becomes a kind of foil for the self-questioning afforded by the game's extended silences, interjecting moments that are simultaneously more like the familiar underscores of action-adventure games and shockingly alienating."[33] As explained by blogger Frederic Fourcade, there is no music during exploration phases of the game. Instead, music is reserved for cut scenes, when approaching monumental locations, and in boss fights. As a result, Fourcade argues:

> Indeed, music is granted a much bigger role than just an accompaniment. It aims to address the lack of explicit narration, no voice, no introduction text, (almost) no word from the hero, nothing tells us who he is, where he comes from, who is the girl he carries, how she died, etc. . . . The music guides the player and helps him imagine how the main protagonist feels, what he went through, and what he hopes for. From this point of view, the compositions of Kow Otani are amazingly powerful; they combine with the visual design of the game, with the framing and with the editing of the cut-scenes, and we realize that the risk taken by Fumito Ueda in his narration presents us with a striking result.[34]

The contrast of silence during exploration versus music during cut scenes arguably heightens the emotional impact of those moments. In one scene cited by players as one of the most impactful of the game, the player-character arrives at a temple after crossing an ancient stone bridge mounted on Agro, their horse and only companion throughout the game. The bridge begins to crumble, and Agro sacrifices herself in order to throw the player-character to safety on the opposite side.

The game has little dialogue, but the music in this scene triggers a strong emotional response from the player, suggesting an instance of existential authenticity.[35] The song "Resurrection" was heard earlier in the game as an extended arrangement in the introductory cut scene and more importantly after defeating the first Colossus enemy and returning to the central temple (where the player sees a woman, Mono, lying on a stone altar awaiting resurrection). In both the Agro and Mono temple scenes, a prominent reverb effect is apparent in the audio, suggesting a large, resonant physical space congruent with the visuals of the stone temple. However, additional structural elements of the music suggest musical practices characteristic of baroque counterpoint that hint at important aspects of the narrative through trope. As shown on figure 2.2, the music features two voices that begin in imitation, duplicating the same pitches an octave apart, but displaced by three beats. The melody also outlines a perfect-fifth interval, a prominent interval in baroque counterpoint, as shown in bars 2–4. Suspensions are frequent, shown with interval numbers in bars

5–8 that indicate the changing harmonic interval between bass and treble pitches; dissonances, as expected in baroque style, resolve down by step on the weaker beat to consonances, producing a tension-and-release effect. And lastly, the piece ends with an accidental raising of the D to a D♯, the raised leading tone characteristic of minor key usage in the baroque and classical eras, particularly at moments of ending or cadence. These features are not entirely surprising for the gamer familiar with the media tropes of castles and temples. As James Cook describes, "frequently, the musical backing for castles and labyrinths either exploited pre-existent baroque music or period pastiche. The associations between baroque music and grandeur may have been partly responsible for this choice of music; baroque architecture and baroque music are often linked in the scholarship and in the popular imagination."[36]

These are the *topics* of this piece, the stylistic features with fixed musical properties that merge to form the *trope*. I have presented this piece to students in my Music of Video Games course, and their observations about the effect of the music on them as players give some insight into how players perceive trope here. Students cited the effect as creating a sense of sacredness or solemnitude, emphasizing mourning or loss, but also evoking sadness and introspection. Obviously their views are colored by the game narrative, which mirrors similar emotions, but there are specific musical properties that contribute to this troping effect of *sacred* as well. The use of musical properties characteristic of seventeenth-century counterpoint evokes the use of counterpoint in sacred works from Western European Christianity such as masses and requiems.[37] The reverb effect also contributes to this association with sacred through its physical location. In combination with the counterpoint and minor mode, the reverb suggests a funeral mass played in a cathedral or large stone church,

Figure 2.2. "Agro Falls" fragment from "Resurrection," *Shadow of the Colossus*. Source: Composed by Kow Otani. Transcription and analysis by the author. Original music copyright Sony Music Artists, Inc.

linking to the stone temple location within the gameworld. Karen Cook describes a similar effect in *The Legend of Zelda: Ocarina of Time*'s Temple of Time, which similarly uses untexted voice emulation with a strong reverb effect, as evoking the ancient and spiritual through both these elements and its associations with plainchant.[38]

This effect is not accidental: the game's composer, Kow Otani, in describing his soundtrack for *Shadow of the Colossus*, states that "as a whole, this soundtrack has more in common with a prayer or a requiem."[39] The requiem, a style that frequently uses counterpoint in commemorating the death of an important personage, itself evokes sadness through the listener's previous experience, and its use here highlights exactly how emotionally important Agro as a companion has been to the protagonist of the game. The emotional descriptors used by my students mirrored the increased complexity of the trope itself. The music does not merely use a superficial reference to something like the *Toccata* to reference horror, but instead combines elements associated with the sacred, particular (visual) physical spaces, and sound effects (rather than simply music) to evoke the trope while removing specific references to Christianity by the music's lack of text to make it more believable within the non-Earth gameworld. Karen Cook discusses this contrast with the modern in her discussion of plainchant in video game. While the *Colossus* example is less plainchant in style and more contrapuntal, I believe her argument still holds here:

> Plainchant [*or counterpoint, in our example*] acts as an exoticism. Its distinct musical style usually marks it as "other" to whatever other music is at play, and as such is often used to denote the "otherness" of a particular character, scene, or world. Yet at the same time, its familiarity acts as a signifier of the past, medieval or not, and of a generic sense of religion if not specifically of Catholicism.[40]

In *Colossus*, these musical elements (topics) merge together to generate the trope/media trope of ancientness depicted in both the narrative and visuals.

Furthermore, the players' understanding of the presented tropes/media tropes can shift throughout the game as they work in conjunction with other components of the narrative. Although the exact same music sounds a second time in the Agro Falls scene as in the earlier Mono Temple scene, players recontextualize their understanding of the music for the new narrative context, perceiving the music as emotionally more powerful here given the attachment they have formed to Agro throughout the game. As Gibbons describes, "while in film and television the audience bears no responsibility, in games, the line between player and avatar is often quite thin; filtered through these alter egos, events are, in a sense, not happening to other people, but to ourselves."[41] An astute player may also recognize that the song, due to its previous association with Mono and her resurrection, hints at a future outcome regarding Agro.

CASE STUDY: MEDIEVALISMS

Thus features of both sound and music can combine with visuals, narrative, and more to generate and reference tropes/media tropes in a more nuanced way. General stylistic features of frequently heard musical styles—what Hatten identifies as topics (although Atkinson rightfully points out that topics and tropes tend to form a continuum rather than a binary opposition)[42]—also play a frequent role in coloring the gameworld. In our last chapter, we discussed how historical elements, albeit somewhat morphed, can inflect gameworld through markers of time and place, narrative inflection, and game-cueing functions. This depiction of gameworld can be generated through topics and tropes, a process that is particularly prominent in the Fantasy genre.

In this section, I will focus on Fantasy-inspired video games, a genre that references the imagined past while incorporating magic or the supernatural. Deriving from Fantasy literature such as Tolkien's *Lord of the Rings* series and stock worlds and characters established in *Dungeons & Dragons*, common media tropes in this genre include imaginary worlds often based on mythical creatures, deposed kings and queens, a rigid dichotomy of good versus evil, adventurer-heroes, and pre-industrial cultures (often modelled on real-life societies).[43] In video gaming, several different styles of gameplay regularly incorporate plot elements from the Fantasy genre: adventure (where the game is structured as a series of quests), role-playing (where the player takes on the role of a single character whose skills improve throughout the game), action (where the game involves fighting via weapons, magic, or hand-to-hand combat), strategy (generally a turn-based setting involving battles/defense or territorial expansion), and massive multiplayer online role-playing games (where players immerse themselves as a particular character in an online setting, interacting with other players across the world). These Fantasy-themed genres seem to share several features such as quest-based narratives, levelling-up and scaled combat, magic as an offensive or defensive force, world exploration (especially in open-world environments), and the acquisition of treasure (loot finding).

The Fantasy genre in modern media may incorporate elements of magic and myth, but it is nevertheless often grounded in a modern-day interpretation of the Middle Ages. Tison Pugh and Angela Jane Weisl identify this as "a framework onto which one can hang various sorts of fantasy. The medieval tropes that succeed within this framework are incorporated, often those that make it look and feel right, although 'right' is primarily defined through a contemporary lens."[44] Research on modern-day medievalisms has blossomed in the past decade in literature and film by scholars such as Pugh and Weisl, Oliver Traxel, Daniel T. Kline, and Robert S. Sturges, but has been slightly later in developing within the study of the music of video games. As Karen Cook explains, "unlike film, the study of medievalist music within video games remains a nascent field, with recent contributions being made by scholars such as [herself and] James Cook."[45] Much of this work has focused on musical tropes, a trend I will expand upon in this chapter, but in the process I will identify how the composers' and game designers' depictions of medieval—that is, their use of *medievalisms*—and the reality of European medieval life and culture do not necessarily correspond, and how this disparity manifests our larger functions for historically referential sound.

What are Medievalisms?

What exactly is a medievalism? Karen Cook, citing Annette Kreutziger-Herr, defines medievalisms as the "reception history of the Western medieval period.... The study of medievalism in music investigates reactions to, reuses or reconstructions of, and creative interpretations of the medieval, both deliberate and inadvertent."[46] James Cook supports this definition and further makes a link between medievalism and Fantasy:

> Medievalism is essentially the creative afterlife of the middle ages. It does not therefore mean "real" medieval music, but musics from other historical periods that are understood "as medieval" in popular culture (from baroque, to folk, to heavy metal), and newly composed works that draw on readily understood symbols of the medieval. It is one of the primary animating forces behind fantasy, and has a rich history in its own right.[47]

Creative interpretations, in my view, is the most important component of both definitions. Medievalisms do not aim for 100 percent historical accuracy, but rather for a suggestion of the Middle Ages that allows enough flexibility to incorporate other modern-day elements. Karen Cook also identifies a key role of musical medievalisms as establishing time and place, but importantly qualifies this as not always evoking a *specific* time and place, stating that "medievalist music also helps to create an atmosphere of antiquity, of unspecified past-ness, that often finds a home in media where a fictitious, even fantastical, setting is portrayed."[48] In video games, much like in television and film, the Fantasy genre is rife with medievalisms, suggesting particular gameworlds that both are and are not reflections of the medieval world. Why do medievalisms feature so strongly in the Fantasy genre? Some such as Daniel T. Kline argue that

Fantasy, with its open-ended narrative structure and quest-based format, is the revival of medieval literary forms such as Arthurian legends.[49] However, since few game developers are experts in medieval studies, the common media tropes of medievalism portrayed through film and video games often are not, in fact, accurately rooted in medieval genres but instead come from a range of sources. Pugh and Weisl cite one gamer who suggests that *medieval* means "anything before the Renaissance, before people were paying too much attention," essentially any period that runs in opposition to modernity.[50] Such an observation is supported by Oliver Traxel, who states that the modern use of medievalisms is "more about evoking a general medieval 'air' than locating the narrative in a particular historical site."[51] These modern medievalisms are nowhere near being historically accurate representations of the Middle Ages as their elements are often limited to the culture of Western Europe, omit the very significant scope and influence of the early (Catholic) Church, and minimize class and political distinctions. I will remind the reader here of Pugh and Weisl's previously highlighted assertion that medievalisms function as a "framework onto which one can hang various sorts of fantasy."[52]

Why does the Fantasy genre suggest the medieval era at all if important elements are jettisoned? Film scholars speculate that medievalism produces effects of desire and nostalgia which modern settings cannot replicate.[53] Pugh and Weisl argue that this faux-medieval world is appealing because it has been sanitized compared to the historical Middle Ages, retaining associations with castles, magic, and weapons but jettisoning other components:

> Fantasy, tautologically yet intrinsically, produces its own world within a system of meanings that overrides the rules of realism while still remaining believable as fantasy, and so the worlds of these games are not simply preposterous pastiches of medieval history, but a version that picks and chooses, creating an authentic atmosphere that allows the game's story to be played out within its constructed framework.[54]

In video games, game designers create their own fictional universes with their own visual styles, sound, music, and even mythology, and thus this constructed framework (analogous to Karen Cook's "creative interpretations") is a vital component. Game scholars often discuss player engagement and immersion in regard to gameworlds. For instance, Collins describes games as "sites of participation and practice where players construct meaning,"[55] while Whalen sees the resultant game immersion as an essential aspect of pleasurability in games.[56] Thus a preexisting framework or set of media tropes gives the game designers a starting point for creating a gameworld's aesthetic design, plot structure, music, and other elements that allow for quick reference to players' previous emotional and aesthetic experiences.

Musical Medievalist Tropes and Functions

Returning to our functional roles for game sound, establishing time and place is very common in medievalist Fantasy games given that many explicitly reference a particular historical event or era. However, Fantasy games that are less explicitly historical still may, as we discussed earlier, use medievalisms to evoke a generalized past (or in some cases a generalized pre-industrial culture). As Karen Cook describes, "medievalist musical tropes are, therefore, often anachronistic, polyvalent, and located at the crossroads of timbre, visual imagery, and plot."[57] Medievalisms in both contexts provide the framework of the gameworld not only musically but also via visuals such as architecture and costume, as well as narrative components such as feudal social structures and the dominance of religion in society. Tropes are a key way of articulating medievalisms, but core to Karen Cook's argument in is that a single trope can sometimes have multiple opposing meanings.[58] As a result, medievalist tropes can go beyond simply establishing time and place, instead also contributing to narrative inflection by suggesting particular states of being or by adding a narrative subtext.

Let us examine how three specific tropes perceived as medieval by players establish gameworld through music: plainchant, folk music, and modalism.

Musical Medievalism #1: Plainchant

> Liturgical plainchant is monophonic and modal, almost always in Latin (the primary exception being the Greek "Kyrie eleison"); it often has a limited pitch range, occasionally contains repeated notes, and can run the spectrum between entirely syllabic and extremely melismatic.... Many chant recordings place the music within a reverberant acoustical space, imitating that of a large church or cathedral. Yet in many of the contexts in which chant has been borrowed or adapted, this list of characteristics is less one of requirement and more one of possibility, wherein any might be manipulated or even ignored.[59]

Several scholars, including Karen Cook and John Haines in particular, have examined the use of musical medievalisms including plainchant. Cook's description above gives an overview of several common features, but more importantly identifies that plainchant in its modern media contexts exists more as an abstract concept than as a historical reproduction. Specifically, creating an effect perceived as plainchant is often more important in modern media than actual plainchant performance. Cook is well aware that game adaptations, while trying to evoke a sense of the past, do not aim for exact accuracy to the past. Her discussion of the wordless voice begins with the observation that "whether singly or in choirs, whether human or synthesized, the voice is widely present, but even when it is linked to the medievalist, it often references Romantic or later practices."[60] Historically, she identifies that "after the Reformation... Plainchant became a signifier, at this point, of Catholicism, oldness or old-fashioned-ness, and ritual, but also of superstition, magic, and perversion,"[61] and that in more recent years "chant in today's popular culture is thus an amalgam, a varying combination of historical plainchant, with its ritual and liturgical overtones, and the resonances of its musical and cultural associations collected over the course of centuries of appropriation."[62] Thus the topic of plainchant and its troping use combine historically accurate elements with other ahistorical references to the genre.

Haines further explains that even our modern attempts to accurately re-create historical plainchant are inaccurate, significantly colored by the plainchant revival in the early twentieth century, and thus do a poor job of recognizing the historical realities of the genre: "This modern assumption about the miraculous sameness of chant for over a millennium has endured to this day, not only in the academic study of chant but, more importantly, in the more common performances and sound recordings of it."[63] Nevertheless, these discrepancies don't seem to matter in modern media given that audiences are unfamiliar with much of the historical context.

What tropes are achieved via the plainchant topic and its associated features? Karen Cook discusses in depth what she terms the "wordless voice" as a common medievalist trope, citing its origins in Gothic literature and film adaptations. As a result, its Gothic subtexts of the supernatural are often a key component of their usage.[64] However, plainchant in particular is used not only to evoke the supernatural but also to signify time periods such as the medieval, or more vaguely the ancient or past, in video games. Additional associated tropes include the aforementioned Christianity, Catholicism, the sacred, and peace and tranquility, as well as more fantastical elements such as the supernatural, magic, evil, and death.[65] Accordingly, several common features emerge from games in this style, although these do not exactly correspond with their medieval counterparts. Text is spiritual and often in Latin, as in medieval plainchant, but does not follow the formal ordering of the liturgy, instead using single words or short phrases for effect. Karen Cook describes fragmented Latin in her own analysis of *Sacred 2: Fallen Angel*:

What is used here is not a complete chant at all, though, but words and melodies excised from the well-known "Pange lingua" chant: "gloriosi," "mysterium," and so forth.... What was important in constructing this moment in the soundtrack was not the Latin text per se, certainly not in any liturgical or grammatical sense, but rather the sound of chant itself; the modal monophony, the sound and color of the men's voices, and the lack of any inherent rhythmic pulse create a musical paradox with the accompaniment that, I argue, enhances the overall narrative of the game and the tension in this particular scenario.[66]

Haines describes the "sound of chant" as closely associated with the style established by the monks of Solesmes in the early twentieth century: low-register male voices, no instrumental accompaniment, and the elimination of vocal idiosyncrasies to produce a smooth, pure tone.[67] While some of these features are common in video game sound (particularly the use of low-register voice or untexted voice reproduction, suggesting monastic chant), this is often paired with non-medieval instruments, including contemporary-style percussion. Similarly, while we will see that some of the upcoming examples begin with monophony, more commonly the chant voices are layered over an orchestral or drone accompaniment. A common feature of historical parallel organum style, perfect-fifth intervals, is also common in video game plainchant, although this is often used more widely to signal the ancient via both melodic and parallel motion between voices (as seen in our *Odyssey* examples last chapter).

Figure 2.3 transcribes a song with many of these features, the opening of *Stronghold 3*'s "A Pane in the Glass."[68] *Stronghold 3* (2011), as a real-time strategy game based in Britain circa 1066 CE, logically uses a plainchant-inspired style to reference a specific time and location within medieval Europe, a feature observed by Karen Cook as common in other historical games such as *Assassin's Creed*.[69] The song, composed by Robert Euvino, begins in unison but transitions after its opening perfect-fourth motive to a melodic voice accompanied by drone, and in the last four notes of the excerpt fills out the texture with a chorale harmonization that is much more typical of the baroque era with four-part harmony and modulations than of the medieval era. This is a phenomenon scholars are well aware of, with researchers such as James Cook describing the "tendency, in recent years, to conflate all preclassical music in popular historical film and television. In effect, baroque music comes to stand for the medieval and Renaissance."[70] As heard in this song, video game chant is fragmentary and recast within new sonic environments. Plainchant-like features in "A Pane in the Glass" include its use of text. Although blurred by a strong digital reverb effect, the opening motive begins with the word "Al-le [lu-ia]," text often associated with sacred Christian chorale settings (although note that this text has not been notated in the given transcrip-

Figure 2.3. Opening, "A Pane in the Glass," *Stronghold 3*. *Source*: Composed by Robert L. Euvino. Transcription and analysis by the author. Courtesy Firefly Studios, Inc.

tion since the reverb blurs its final syllables too much to place them precisely here). The slow tempo, Aeolian/natural minor mode, and reverb also evoke players' medievalist stereotypes of plainchant sung in cavernous cathedrals. In addition to the previously described analyses by Karen Cook, many of these features are also identified by James Cook in his own analysis of a newly composed medievalist passage in *Civilization V*. He cites the use of male choir, Latin text, voices paired at the fifth, Dorian mode, and open-fifth drones in that work as a combination that "immediately situates us within a medievalist sound world."[71] These features allows players to quickly recognize the topic components and connect them to their associated medievalism trope.

Interestingly, the games that reference precise historical dates and locations at times feature the strongest sonic recontextualization. *Medieval II: Total War* (2006), a turn-based strategy game set between the years 1080 and 1530, is one such example.[72] "Amen," composed by Jeff van Dyck, begins with a tabla-like hand drum, a low-register plucked-string accompaniment, and additional swelling sound effects, but is soon joined by fragments of unison male vocal plainchant set to the text "Alleluia . . . Amen . . . Hosanna."[73] Some of its musical features as shown in figure 2.4 are typical of medieval plainchant; the melody is in Aeolian mode with prominent perfect-fifth leaps and whole-tone motion to scale-degree 1.[74] However, the Latin text is once again fragmented, consisting of single words from the Catholic liturgy rather than the full text. And at 1:05, voices move in parallel fifths, followed shortly by organ and digitally reproduced recorders in a contrapuntal style more reminiscent of baroque church keyboard styles. The blend is anachronistic in multiple ways, with its elements evoking a generalized past but not exclusive to the Middle Ages or to Europe, a reference that Karen Cook terms "a general sense of oldness."[75]

"Acre—Underworld," composed for *Assassin's Creed* (2007) by Jesper Kyd, also uses a chant-like excerpt (figure 2.5) while placing it in a decidedly non-medieval context despite the gameworld setting of the Middle East during the Third Crusade (1191 CE).[76] Modern instruments, including a heavy bass line, digitally altered sounds, and piano, are evident from the beginning of the excerpt.[77] Plainchant appears in fragments, with short excerpts articulating the text "Exaudi Nos," "Agnus Dei," and "Miserere Nobis" (the chant is, in fact, sampled from a preexisting recording). The

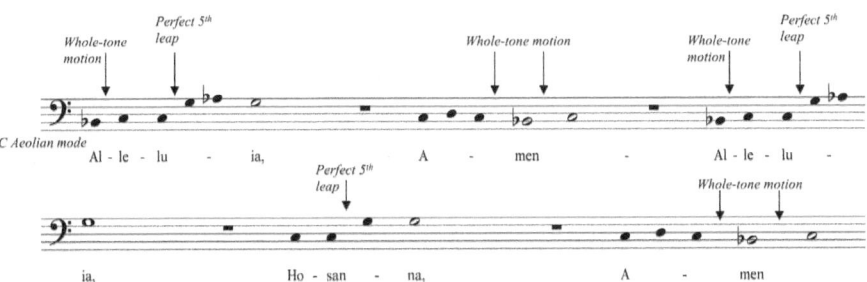

Figure 2.4. Plainchant fragments at the beginning of "Amen," *Medieval II: Total War*. *Source*: Composed by Jeff van Dyck. Plainchant fragments from Catholic Church, *The Parish Book of Chant* (Richmond, VA: Church Music Association of America, 2008). Figure and analysis by the author.

vocal line features a repeated recitation tone that reflects the rhythm of the spoken text, both features characteristic of medieval plainchant. In this game, the combination of modern and medievalist elements is intentional, mirroring the plot structure of the game wherein the protagonist bridges the twelfth-century Holy Land and the modern era via technology.

In all three examples, plainchant acts as a signifier of the medieval, but it is not heard in isolation. Its signifying function is one of *blending* medieval with more modern musical elements, establishing time and place as a hybrid of the gameworld and the digital medium. As Karen Cook describes, "new media genres such as film, television, and video games thus simultaneously play with the resonance of chant's previous associations and add to its symbolic tool chest . . . the idea of chant in modern media is not bound to liturgical or historical plainchant; it is omnivorous, accepting music that samples, remixes, and emulates plainchant in myriad ways."[78] Furthermore, plainchant, organ, and baroque counterpoint and chorale harmonization are all used to signify the historical sacred more broadly rather than maintaining strict adherence to stylistic norms of the medieval era.[79] The effect of oldness is much more important in establishing a true-to-expectations experience of authenticity than is strict adherence to historical accuracy.

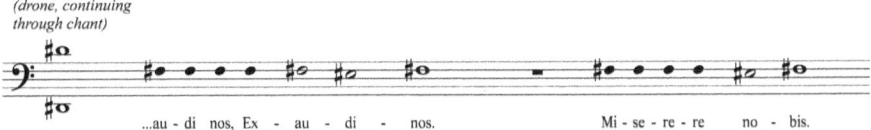

Figure 2.5. Plainchant excerpts in "Acre—Underworld," *Assassin's Creed 1*. *Source*: Composed by Jesper Kyd. Transcription and analysis by the author. Reproduced by permission of Ubisoft Entertainment SA and Hal Leonard, Inc.

Musical Medievalism #2: Folk Music

Medievalist musical borrowings are not limited to sacred music. Perhaps due to the common inclusion of bard characters in Fantasy,[80] there is a prominent use of Western European folk music styles and instruments. Scholars such as Karen Cook, John Haines, and Elizabeth Randall Upton have noted the particular emphasis on folk music traditions from the United Kingdom and Ireland in the Fantasy genre.[81] Such usage has parallels in film and, as James Cook identifies, can be problematic in its deceptive representation of these cultures:

> Celtic folk music as a shorthand for the Middle Ages is nothing new. Simon Nugent has noted the tendency of many historically-situated films to draw on Celtic influence. His work has shown that the creation of "Celtic" folk has little to do with a discrete geographical area or with historical accuracy but rather is a modern marketing creation that plays on associations with nature and an escape from modernity.[82]

What is the value in medievalist references that are not actually medieval? As James Cook cites from Kim Selling, such references act as narrative shorthands, allowing the viewer to pull from their previous knowledge of both history (what real-world setting is the game alluding to?) and Fantasy (what do its associated media tropes and archetypes inform us about the narrative in this game?).[83] In order for players to access these references, they must be simple enough to be recognizable to the player. While audiences may not know or recognize the distinction between, for example, specific Irish and Scottish fiddle ornamentations, they will make a connection between the genre as a whole and the British Isles. And in doing so, they will refer intertextually to their knowledge of pre-industrial depictions of these cultures.

Such, I would argue, is the reason behind the inclusion of preexisting British folk songs in *Stronghold 3*. Songs such as "Tom of Bedlam" and "John Barleycorn" allude to the game's setting in 1066 CE Britain. However, anachronisms in both the choice of songs and instrumentation suggest that the game developers were more concerned with evoking a general sense of folk style rather than historical and geographical precision. Specifically, the renditions of these two songs feature instruments common in modern-day Irish folk song performance, such as fiddles, whistles, bodhrán, and guitar, rather than period-appropriate instruments. As Haines describes, "three instruments especially, the bagpipe, bodhrán and pennywhistle, became closely associated with contemporary Celtic music in film and elsewhere. Just as actors' vaguely Scottish accents have been popular in medieval films of the last two decades or so (with the post-Bond Sean Connery leading the way), these musical instruments are indispensable musical signifiers of the Middle Ages."[84] The two songs themselves are still performed by modern-day folk groups, making them part of a *living* tradition rather than historical artifacts. But as Randall Upton observes, the genre invokes associations with an older, simpler time in modern media: "Folk music reads as traditional, as old, but not ancient. The folk music sounds inhabit an intermediate place, a remembered past between the present and the imagined distant past of the middle ages."[85] Although it is not possible to exactly date these folk songs given their oral transmission, the texts suggests that the songs are newer than the *Stronghold 3*'s 1066 CE historical setting, with both John Barleycorn and Tom of Bedlam being popular characters in poems and songs of the seventeenth to eighteenth century.[86] The songs are old, but not from the turn of the first millennium. As Haines describes, "the idea of an implicit connection between modern folk music and medieval music, created in the sixteenth century and seldom questioned since, has had a strong influence on medieval cinema" and other forms of media.[87]

The use of musical anachronisms and more creative modifications to the source materials are also common in adapting older songs to modern-day aesthetics. This historical alteration is, on the whole, fully accepted by fans and nonacademic media. For example, Luke Plunkett, a writer for game culture website Kotaku, admires that the sea shanties in *Assassin's Creed: Black Flag* texture the gameworld by filling in details of moment-to-moment experiences, but he admits that "most sea shanties you recognise from *Black Flag*, including the famous 'Drunken Sailor,' were actually first sung decades, sometimes even centuries later."[88] Plunkett further details the history of several sea shanties within the game, giving examples of how songs were repurposed from other genres and eras and had their lyrics sanitized for modern audiences (such as the removal of the N-word from "Johnny Boker," a shanty repurposed from a nineteenth-century minstrel song by J. W. Sweeney). Such creative adaptations are commonplace and expected yet do not weaken the player's sense of authenticity; indeed, retaining language no longer considered socially acceptable would be much more jarring from a player perspective. The compromise is made to create a stronger sense of immersive gameworld for the player.

Direct use of preexisting folk songs is not the only way that this style is used. Equally common is the use of newly composed music that imitates folk styles with similar instrumentation, meter, mode, and rhythms.[89] Take, for example, "Moldheart's Hornpipe" from *Sims Medieval* (figure 2.6), composed by John Debney.[90] The instrumentation is suggestive of a traditional Celtic folk ensemble, with whistle/pipe, folk dulcimer, harp, fiddle (at times functioning as a hurdy-gurdy drone), tambourine, and a hand drum.[91] The whistle's melodic idioms evoke folk-reminiscent improvised ornamentation, with mordents and trills on longer-duration

Figure 2.6. Bars 1–20, "Moldheart's Hornpipe," *Sims Medieval*. Source: Composed by John Debney. Reduction and analysis by the author. Reproduced by permission of Electronic Arts and Hal Leonard, Inc.

and downbeat notes, and the dulcimer and harp accompaniments focus strongly on drone tones on the tonic and dominant (F and C in our current key of F major). Much like in our plainchant and Greek-inspired example, drones are often used to evoke the musically ancient. In conjunction with this drone, there is very little harmonic change in the first sixteen bars of the music, with a constant reiteration of F and C to suggest tonic harmony, although the whistle at times plays against the resounding F major harmony with longer durations on G (bars 2, 3, 6, 9, 10, and 13) and D (bars 4, 7, and 12, an addition which suggests a shift to the submediant chord). Finally, the hand percussion suggests an improvisatory style since there is often variation from one bar to the next and one phrase to the next; for example, the tambourine in bars 9–12 is a varied repeat of bars 5–8, subdividing longer durations or adding a roll to increase the rhythmic density. This combination of improvisatory style, drone accompaniment, and an ornamented, fast-moving whistle melody accompanied by folk instruments creates a medievalist trope because players understand folk music as equating to an older musical tradition. As James Cook states, "this 'packaging' brings with it associations of an 'authentic' Celtic folk tradition as a remnant of the 'true folk tradition' that once existed for everyday people elsewhere."[92]

In bar 17 of the example, a new section begins and the instrumentation changes significantly. Whistle, fiddle, and dulcimer now play in unison in a lower octave than the melody of the previous section, and the whistle's ornaments are notably absent. Low-register bowed strings replace the previous dulcimer and harp accompaniment with a substantial acceleration in the rate of harmonic change (now one chord per dotted quarter duration). This is paired with significantly denser percussion, with the tambourine playing every eighth note and the hand drum repeating a short-short-short-long pattern in shifting combinations. The overall increase in density (eight simultaneous notes rather than five or six), rhythm, and lower overall register creates a feeling of energy and drive for the player, but subtle inconsistencies occur in the folk style; bowed string chords, for example, are not particularly characteristic of Celtic folk ensembles, and the whistle is playing in a lower range that would not normally be playable without switching to a second instrument. However, players do not notice this change of accompaniment and range as a break in authenticity, first because the acceleration and drive of this new section take priority in their listening, and second, because many players are not familiar with Celtic folk music beyond their previous media experiences. In this case, game *function* takes over in the player's perception, shifting the importance to constructive authenticity (player expectation) and existential authenticity (player's emotional responses, including tension and anticipation in this case) over objective authenticity (the historically verifiable components of the music).

Musical Medievalism #3: Modes and Cadence

I discussed in our last chapter on gameworld that the diatonic modes are often used as a symbol of the musically ancient. Anything prior to the age of industrialization may be perceived as ancient by players, and thus a signifier used to represent Ancient Greece circa the fifth century BC can (and *is*) equally used to evoke the medieval era. Fantasy games, much like more literally historical games, also make use of this signifier, but unlike our Ancient Greek–inspired music and true historic medieval music, these tend towards a much more limited selection, relying primarily on Aeolian and Dorian modes in addition to the major scale (Ionian mode) while less frequently using Phrygian, Lydian, or Mixolydian modes.

Why is this the case? Partly, I believe this occurs because Aeolian and Dorian are the two modes that most closely resemble the minor scale. Aeolian duplicates the natural minor scale, while Dorian has a raised scale-degree 6 in comparison with natural minor. Notably, they both contain a scale-degree 7 that is a whole tone away from the tonic, in opposition to the semitone heard from scale-degrees 7–1 in major, and they also support minor chords on scale-degrees 1 and 5, two musical features that are often used as signifiers of medieval sonorities in video game music. A few examples will help to illustrate these features as musical signifiers.

The song "Age of Oppression"/"Age of Aggression" from *The Elder Scrolls V: Skyrim* is a typical example of tavern bard song in Fantasy games.[93] It is notably in D Aeolian (natural minor) and cadences at the end of each four-bar phrase with a longer duration note followed by rests. Each cadence uses one of the four patterns shown in figure 2.7: a leap from A to D (scale-degrees 5–1), step motion to D from above or below (scale-degrees 2–1 or 7–1), or a leap from C to A (scale degrees 7–5, suggesting a half cadence). The emphasis on step-motion (whole-tone) cadence patterns occurs frequently in Fantasy video game music, combining with modes (and the texted bard song in this example) to produce a medievalist trope. These cadences notably *avoid* any leading-tone (semitone) motion to its tonic scale-degree, a feature emblematic of historically more recent styles ranging from the Renaissance and baroque through modern-day pop and jazz music. In "Age of Oppression"/"Age of Aggression," the musical trope is further reinforced by visual and narrative

Figure 2.7.   Cadence patterns in "Age of Oppression"/"Age of Aggression," *Skyrim*. *Source*: Figure by the author.

elements; in-game bards play the lyre and wear costume reminiscent of the middle ages, for instance.[94] Bard songs in *Skyrim* feature an emphasis on narrative verse set to strophic melodies, through which bards act as storytellers in gameworld social spaces such as taverns. The lyrics of "Age of Oppression"/"Age of Aggression" and other bard songs comment upon the factional division between the Stormcloaks and the Empire within the gameworld, and vary depending on whether the player is in an area controlled by one faction or the other within the game.

The whole-tone cadence pattern also occurred in several of the previously discussed examples; the B♭ to C motion in *Medieval 2*'s "Amen" (our previous figure 2.4), for example, as well as the E-to-D and G-to-F motion in "Moldheart's Hornpipe" (figure 2.6, which despite being in the major mode achieves its whole-tone cadences through motion from scale-degrees 2-to-1 or 7-to-6). While this pattern occurs diatonically from scale-degrees 7-to-1 in Dorian, Aeolian, and Phrygian modes, as Carl Dahlhaus observes in his *Studies in the Origin of Harmonic Tonality*, as early as the thirteenth century the leading-tone cadence pattern began to be preferred, even in these modes.[95] Game composers therefore employ this whole-step cadence much more frequently than might be expected in true medieval style. However, if game composers were to employ the more common leading-tone (semitone) cadence pattern in these modes, which does not differ from modern usage in minor-mode tonality, such an effect would not distinguish the resolution as "musically ancient" to the casual listener since semitone cadences are heard as normal to the modern gamer and thus do not provide a signifier of the medieval. Whole-tone cadences are less common and thus more noticeable as a signifier of otherness.

Other ways of avoiding semitone motion from 7-to-1 also occur in this repertoire. Another song from *Skyrim*, "Out of the Cold" (figure 2.8), uses a similar whole tone cadence in the key of E-flat major at the end of its first eight-bar melodic phrase from F to E-flat, but later phrase endings shift this whole tone to different scale-degrees (example [a] on the figure), vary the cadence through the repetition of the cadence pitch to signal an ending

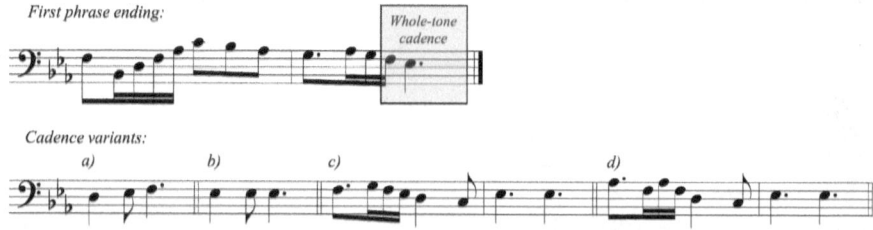

Figure 2.8. Cadences in "Out of the Cold," *Skyrim.* **Source: Figure by the author.**

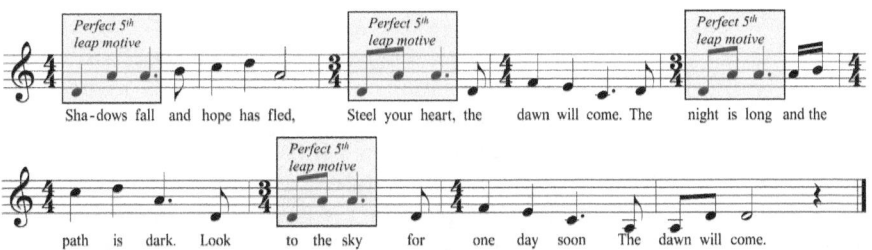

Figure 2.9. Perfect-fifth leaps in "The Dawn Will Come," *DragonAge: Inquisition*. *Source*: Composed by Trevor Morris. Reduction and analysis by the author. Courtesy Electronic Arts.

(b), or employ the 7–6–1 scalar motion characteristic of the so-called *Landini cadence* of the medieval era's fourteenth to fifteenth centuries (c and d).

Additional melodic and rhythmic features are paired with modes and cadence in such music to strengthen the signifiers of medievalism even when games do not reference the historical Middle Ages. *DragonAge: Inquisition*, an action/RPG game released in 2014 with music by Trevor Morris and Raney Shockne, is one such example.[96] "The Dawn Will Come" (figure 2.9) features a strong emphasis on the perfect-fifth interval, as seen in the opening D-to-A melodic leap that recurs every two bars. Changing meter in this example is used to create the effect of unmetered plainchant (although the repetition of the D-A motive in the same [eighth, eighth, dotted-quarter] rhythm undermines this somewhat by introducing an element of regularity that suggests a downbeat). The song's use of Dorian mode, its avoidance of semitone cadences, its lyrics, and the narrative context of the game (where the song is used as a prayer to rally allies by the priestess Vivienne) make the trope and media trope clear: The song is a representation of the game's religion, employing real-life signifiers of the sacred and ancient despite the game's fictional gameworld and triggering intertextual references to the sacred as mystical to support the game's Fantasy genre. The trope is not created by merely one of these musical features, but rather the ensemble of the whole.

TROPES: REALITY VERSUS FANTASY?

So why, if attempting to create a medieval style, would game designers and composers not use actual medieval styles and genres? In Traxel's words, "there are also truly pseudo-Medieval elements in these games, elements that are presented as Medieval yet have no basis in the history or literature of the Middle Ages."[97] He suggests some of this is due to accessibility reasons, such as translating Latin text (which most modern gamers cannot read), but also argues that game designers are looking for louder, faster-paced music to create energy and excitement within the game. Additionally, Traxel argues that authenticity is hard to produce given that the Middle Ages encompassed a large period of time and thus a range of musical styles, and therefore the game designers prefer a generalization of features considered medieval. He suggests a better motivation for the use of medievalist elements is that "the historical Middle Ages have enough drama associated with them, yet enough ambiguity, to serve as an ideal backdrop for adventure, particularly for those games involving the combat with which the Middle Ages are often connected. Moreover, the Middle Ages have enough continuity with modern Western culture to be somewhat accessible, yet are distant enough from that culture to also be rather exotic."[98]

While I agree with most of these points, and particularly that the game designers are looking for a model of pre-industrial society as a scaffold for their Fantasy/magical elements, I believe the use of medievalisms goes beyond Traxel's explanations. Quite simply, I believe that the game designers are looking for a level of familiarity among their audience. Since their audience is *unfamiliar* with true medieval musical style, features that distinguish this repertoire from modern music, such as whole-tone cadences, modes, and plainchant, are an easier method by which to accomplish historical referencing. Indeed, in cases where the game designers and composers attempt to integrate modern and medieval themes, such as that presented in *Assassin's Creed*, there needs to be a clear distinction between "what sounds modern" and "what sounds medieval" in order to effectively combine the two styles. In other games this distinction is not quite as important, but media tropes of bards, sacred mysticism, and Celtic (or in the case of *Skyrim*, Nordic) inspired cultures can help quickly set the tone of a particular game environment through preexisting cultural associations.[99] Game designers are ultimately interested in creating gameworlds that are as immersive as possible in order to have players spend more time with the product; these medievalisms and their associated media tropes quickly establish the game setting and mood through reference to the players' and designers' previous knowledge.

However, are there any problems in utilizing this fake version of history? As I discussed last chapter, many scholars say yes—by presenting players with a false version of history, players learn the wrong version of history.[100] Our medievalist tropes were rife with incorrect dates for particular musical styles, instruments drawn from completely different environments than the Celtic-inspired cultures of the game, and the assumption that folk music is an ancient and unchanging tradition (which couldn't be further from the truth in respect to Celtic folk styles). As James Cook explains, "music can historicize in the popular imagination in a manner not always indebted to temporal reality."[101] Players bring these incorrect assumptions back into the real world.

Because tropes are often kept simple to understand and are understood intuitively through memory, they have a direct impact on nostalgia, game immersion, flow, and the perception of recurring thematic ideas. However, this simplicity can also work against tropes. In the desire to quickly communicate information, tropes can flatten many characteristics of cultural references, leaving them overly simplified, removing nuance, and sometimes veering into problematic representations. Our examples thus far flattened out many features of Celtic folk music and medieval sacred traditions, but this occurs outside of medievalisms as well. Take, for example, the game *Samba de Amigo*, originally released for arcade consoles and the Sega Dreamcast game console in 1999.[102] This game, in which the player shakes maraca controllers in sync with the music, attempts to create a party-like experience through features such as fast music and background characters that dance wearing a variety of brightly colored costumes. Running with the novelty theme of maracas as a video game mechanic, several visual media tropes try to establish a Latin American feel, including the sombrero-wearing main character (a dancing monkey) and a scantily clad background dancer in a butterfly costume suggestive of Brazilian carnival dancers, as seen in figure 2.10.

Figure 2.10.   Screenshot, *Samba de Amigo.* **Source: Reproduced by permission of Sega.**

The references in the game mix cultural elements from an extremely wide geographic area (Mexico, the Caribbean, Central America, and South America) while implying that they represent a single pan–Latin American culture. Thomas B. Yee notes a similar pan-Latin approach in *Super Mario Odyssey*'s Tostarena game level, explaining:

> Pre-release box art and early trailer footage showcase Mario donning a sombrero and poncho in the Mexico-inspired Tostarena. The town's colorful architecture, nopales, and calacas with sombreros, ponchos, and maracas are clear references to Mexican culture. Now while some Latinx fans praised the inclusion of Mexican elements, others objected to the depiction as stereotyping and cultural appropriation. Exasperation was specifically directed at Mario's sombrero poncho outfit which critics described as a caricature of a stereotype and like a thoughtless college student at Halloween. Sombreros and ponchos have long histories in media as caricatures of Mexican culture, making their ubiquity in Tostarena especially egregious . . . [musically, the level's focus on Mariachi style signalled] a sure-fire means to musically encode Mexican identity. The stereotype flattens multi-faceted Mexican culture into an aural calling card, just as conflating Mexican and Cuban elements reduces distinct Latinx cultures into a pan-Latin conglomerate.[103]

Yee terms this usage of real-world cultural elements to create fictional works *racialized fantasy*, and explains that it is problematic both because of the cultural flattening and also because of the focus on so-called *authentic* elements: "Truly authentic elements may be fixated upon and commodified as markers of racial difference. . . . Just as authenticity can be co-opted in dynamics of power, commercialization, and appropriation, music external to a style may be adapted for rhetorical ends."[104] While the problems of representation here are multifaceted, they also highlight the dilemma of players and their interactions with the various types of authenticity. Specifically, individual characteristics that might under other circumstances be identified as objectively authentic are presented out of their original context and too often merged to signify new pan-cultural identities, fetishizing the disembodied symbols of cultures over more complicated cultural contexts. From a constructive authenticity standpoint, players come to expect these cultural flattenings, reinforcing and perpetuating inappropriate stereotypes. And from an existential authenticity perspective, while equating a culture to a particular game dynamic or sense of fun might seem innocent at first glance, it risks ignoring (or replacing) more serious realities and nuance within the greater lived experience of these cultures.

Such issues were not at the forefront of audience's minds upon *Samba de Amigo*'s release in 1999. Pan-Latinism is a frequent approach to the game's musical design, a combination of songs by Latin American artists such as

Ricky Martin, popular songs with references to Latin styles such as Quincy Jones's "Soul Bossa Nova," and non-Latin popular songs rearranged in pseudo-Latin style (not necessarily adhering to a single Latin American musical style, but instead achieved through the use of heavy percussion rhythms, castanets, and particularly instrumentation such as added trumpets). For example, the game's cover of "Tubthumping," a song originally released in 1997 by British anarchist rock band Chumbawamba, speeds up the tempo from the original 103 bpm to a frantic 134 bpm and adds additional hi-hat while increasing the volume of the drums relative to the rest of the mix.[105] The game's cover of "Take on Me" recasts the song's original synth-heavy 1980s pop style to a grunge-inspired hard rock style but unstylistically replaces keyboard with trumpet.[106] Trumpet and percussion thus, in combination with changes of tempo, blend to create the stereotyped trope/media trope of Latin despite the two songs having no Latin American origin. While this passed without much pushback from audiences at its original release in 1999, modern sensibilities would likely find this unacceptable today. As James Buhler describes, "scholars sensitive to race and the dynamics of colonialism have offered an especially cogent critical reception of the theory of musical topics. As scholars of film music have long noted, scores have deployed musical topics to gain clarity in signification but at the cost of resorting to and reifying pernicious stereotypes."[107] A similar critique occurs with the use of medievalisms. As James Cook explains in his analysis of music from *Civilization V*, music can unintentionally act as a tool of Western cultural imperialism, positioning non-Western cultures and non-modern time periods as "other." This sort of musical representation "draws on a pervasive sense of exoticism whereby music of the distant past is seen to be discernible in music of cultures geographically distant from Western civilization. This reinforces a major criticism of the game series, that Western civilization is seen as the natural culmination of progress."[108] Karen Cook similarly notes that such depictions of progress are often very Western oriented and show cultural developments over time in ways that are not accurate to history, and in some cases have soundtracks that "thus narrate a particularly American concept of elitist culture."[109]

CONCLUSION

Culturally, we have particular expectations created based on our previous storytelling (and story-listening) perspectives. From our earliest childhood fairy tales, we learn media tropes as key signifiers of narrative ideas: from fairy godmothers and underdog protagonists triumphing over evil to more modern representations of angst-filled superheroes, tropes not only pervade the media we watch but also create a framework of expectation for narrative events, gameworld, and music. With the advent of video games and a new global media culture, the use of topics and tropes as described in this chapter creates a rich sea of references beyond a single work of media, examining but also exploiting motifs that occur culturally and cross-culturally. Players are thus pulled into a shared culture created through rich references to their lived experiences with social media, online gaming, Netflix, and more. The use of tropes such as those described in this chapter color our experiences of the game at the same time as we bring our game experiences back into the real world, creating new, exciting, and engaging modes of multimedia storytelling, but with a reflexive conundrum. While depicting narrative ideas musically can be particularly effective in storytelling, it is clear that there is a significant gap between the real world and our gameworld fantasies. As we move into other modes of representation in the twenty-first century, both technological and artistic, these are significant issues that need to be addressed.

NOTES

1. Whalen and Taylor, *Playing the Past*, 26.
2. Gibbons, *Unlimited Replays*, 23.
3. Neumeyer and Buhler, *Meaning and Interpretation of Music in Cinema*, 184.
4. van Elferen, "Analysing Game Musical Immersion."
5. Atkinson, "Soaring Through the Sky"; Summers, "Epic Texturing in the First-Person Shooter"; Karen M. Cook, "Medievalism and Emotions in Video Game Music," *Postmedieval* 10, no. 4 (2019): 482–497. Atkinson also cites several notable presentations in his own article, including those by William Ayers, Wesley Bradford, and Marina Gallagher.
6. For example, the historical term *trope* in medieval music refers to insertions of new content to preexisting liturgical chant. That definition will not be applicable here.
7. Richard Nordquist, "The Four Master Tropes in Rhetoric," *ThoughtCo.*, accessed August 11, 2021, https://www.thoughtco.com/master-tropes-rhetoric-1691303.
8. Danuta Mirka, "Introduction," in *The Oxford Handbook of Topic Theory*, ed. Danuta Mirka (New York: Oxford University Press, 2014), 24. Mirka's introduction gives an excellent summary of historical foundations in the eighteeenth century for our present-day development of topic theory and also clearly summarizes developments in the field since 1980.
9. Robert Hatten, "The Troping of Topics in Mozart's Instrumental Works," in *The Oxford Handbook of Topic Theory*, ed. Danuta Mirka, 514.
10. "Tropes," *TVtropes*, accessed June 8, 2020, https://tvtropes.org/pmwiki/pmwiki.php/Main/Tropes.
11. See, for example, Dana Plank, "From the Concert Hall to the Console: Three 8-Bit Translations of the *Toccata and Fugue in D Minor*," *Bach* 50, no. 1 (2019): 32; James Cook, "Game Music and History," in *The Cambridge Companion to Video*

Game Music, eds. Melanie Fritsch and Tim Summers (Cambridge: Cambridge University Press, 2021), 344. Karen Cook also provides examples of medievalist tropes in Berlioz and Lully in "Medievalism and Emotions in Video Game Music," 491.

12. Neumeyer and Buhler, *Meaning and Interpretation in the Music of Cinema*, 186; TVtropes, "Trope."

13. "Useful Notes/Bollywood," *TVtropes*, accessed June 8, 2020, https://tvtropes.org/pmwiki/pmwiki.php/UsefulNotes/Bollywood.

14. "Mary Sue," *TVtropes*, accessed August 4, 2021, https://tvtropes.org/pmwiki/pmwiki.php/Main/MarySue.

15. See Neumeyer and Buhler, *Meaning and Interpretation of Music in Cinema*; James Buhler, *Theories of the Soundtrack* (Oxford: Oxford University Press, 2018); and Robert S. Hatten, *Interpreting Musical Gestures, Topics, and Tropes: Mozart, Beethoven, Schubert* (Bloomington: Indiana University Press, 2004).

16. Neumeyer and Buhler, *Meaning and Interpretation of Music in Cinema*, 184–185. They further state "a troping effect is laid on a topic when it is treated in an expressive rather than neutral fashion, an effect that is cumulative. Thus, a single altered element (such as guitar rather than piano, added reverb, slower or faster than expected tempo, or a distinctive articulation such as *non legato*) may or may not significantly alter the topic, but more than one . . . almost certainly will. An analogous topical effect is laid on a trope by repetition, as in the establishment of motifs (whether visual, verbal, or aural)."

17. Ibid., 185.

18. Royal Brown 1994, 239, as cited in Neumeyer and Buhler, *Meaning and Interpretation in the Music of Cinema*, 186.

19. Atkinson, "Soaring Through the Sky," 5.

20. Ibid., 19.

21. John Haines, *Music in Films on the Middle Ages: Authenticity vs. Fantasy* (New York: Routledge, 2014), 45.

22. Hart, "Semiotics in Game Music," 223.

23. Johann Sebastian Bach, *Toccata and Fugue in D Minor*, BWV 565, ed. Wilhelm Rust, *Bach-Gesellschaft Ausgabe*, band 15 (Leipzig: Breitkopf und Härtel, 1867).

24. Neumeyer and Buhler, *Meaning and Interpretation*, 186. This is further discussed in James M. Doering, "Status, Standards, and Stereotypes: J. S. Bach's Presence in the Silent Era," *Bach* 50, no. 1 (2019): 5–31, an article that provides a larger context into the use of J. S. Bach's (and other Bachs') music in the era of silent film.

25. Plank, "From the Concert Hall to the Console," 32–33.

26. Ibid., 34–35.

27. Gibbons, *Unlimited Replays*, 3.

28. *Dark Castle* (Silicon Beach Software, 1986), Mac, PC, Commodore 64, Amiga, Atari, IIGS, Genesis, CD-I, and MSX, sound by Eric Zocher. Gameplay can be found at https://youtu.be/lpJrvAhTiOw.

29. Plank, "From the Concert Hall to the Console," 42.

30. Ibid., 46.

31. *Shadow of the Colossus* (Sony Computer Entertainment, 2005), PlayStation 2, music by Kow Otani.

32. William Gibbons, "Wandering Tonalities: Silence, Sound, and Morality in *Shadow of the Colossus*," in *Music in Video Games: Studying Play*, eds. K. J. Donnelly, William Gibbons, and Neil William Lerner (New York: Routledge, 2014), 122–137.

33. Ibid., 129.

34. Frederic Fourcade, "The Art of *Shadow of the Colossus* (5/6): Music," *Game Developer* (formerly *Gamasutra*), accessed June 29, 2020, https://www.gamedeveloper.com/audio/the-art-of-shadow-of-the-colossus-5-6-music.

35. The "Agro Falls" scene can be viewed at https://youtu.be/O9v8G-68wfs.

36. James Cook, "Game Music and History," 348–349.

37. I will mention here that *sacred* as a concept is very broad; the connection with Western European Christianity discussed here is only one specific instance of how *sacred* might be evoked and is not meant to limit the term's applicability to only that particular cultural context.

38. Karen Cook, "Beyond (the) *Halo*," 197.

39. Dais Kawaguchi, "*Shadow of the Colossus* Composer Interview, iam8bit 2-LP Vinyl Revealed," *PlayStation.Blog*, accessed July 6, 2020, https://blog.us.playstation.com/2018/01/25/shadow-of-the-colossus-composer-interview-iam8bit-2-lp-vinyl-revealed/.

40. Karen Cook, "Beyond (the) *Halo*," 190.

41. Gibbons, "Wandering Tonalities," 127–128.

42. Atkinson, "Soaring Through the Sky," 22.

43. James Cook cites the influence of *Dungeons & Dragons*, sword-and-sorcery fantasy, and the works of Tolkien in particular on the genre ("Game Music and History," 349).

44. Pugh and Weisl, *Medievalisms*, 127.

45. Karen Cook, "Medievalism and Emotions," 482.

46. Ibid., 483.

47. James Cook, "Game Music and History,," 345.

48. Karen Cook, "Medievalism and Emotions," 484.

49. Daniel T. Kline, as quoted in John Gravois, "Knights of the Faculty Lounge," *The Chronicle of Higher Education* 53, no. 44 (July 6, 2007).

50. Pugh and Weisl, *Medievalisms*, 126.

51. Oliver Traxel, "Medieval and Pseudo-Medieval Elements in Computer Role-Playing Games: Use and Interactivity," in *Medievalism in Technology Old and New*, eds. Karl Fugelso and Carol L. Robinson (Cambridge: D.S. Brewer, 2008), 131.

52. Traxel, "Pseudo-Medieval Elements," 130–131; Pugh and Weisl, *Medievalisms*, 108 and 127.

53. Robert S. Sturges, "Medievalism and Periodization in *Frozen River* and *The Second Shepherds' Play*: Environment, Class, Miracle," in *Medieval Afterlives in Popular Culture*, eds. Gail Ashton and Daniel T. Kline (New York: Palgrave Macmillan, 2012), 86.

54. Pugh and Weisl, *Medievalisms*, 126–127.

55. Collins, *Playing with Sound*, ix.

56. Zach Whalen, "Play Along: An Approach to Videogame Music," *Game Studies* 4, no. 1 (November 2004): paragraph 3, http://www.gamestudies.org/0401/whalen/.

57. Karen Cook, "Medievalism and Emotions," 484.

58. Ibid., 482–497.

59. Karen Cook, "Beyond (the) *Halo*," 188.

60. Ibid., 491.

61. Ibid., 186.
62. Ibid., 188.
63. Haines, *Music in Films on the Middle Ages*, 114.
64. Ibid., 490–493.
65. Karen Cook, "Beyond (the) *Halo*," 191.
66. Ibid., 189.
67. Haines, *Music in Films on the Middle* Ages, 114.
68. *Stronghold 3* (7Sixty, 2011), Microsoft Windows, MacOS X, and Linux, music by Robert L. Euvino. The full song may be heard at https://soundcloud.com/firefly-studios/a-pane-in-the-glass-stronghold.
69. Karen Cook, "Beyond (the) *Halo*," 191.
70. James Cook, "Game Music and History," 350.
71. Ibid., 354.
72. *Medieval II: Total War* (Sega, 2006), Microsoft Windows, MacOS, Linux, Android, and iOS, music by Jeff van Dyck, Richard Vaughan, and James Vincent.
73. A recording of the work can be found at https://www.last.fm/music/Jeff+van+Dyck/Medieval+II:+Total+War.
74. While the third of the mode is not included in the excerpts on the figure (and thus the mode could potentially be Aeolian or Phrygian), the overall setting of the song does include E♭.
75. Karen Cook, "Beyond (the) *Halo*," 189.
76. *Assassin's Creed* (Ubisoft Entertainment, 2007), PlayStation 3, Xbox 360, and Microsoft Windows, music by Jesper Kyd.
77. A recording of the work is available at https://www.youtube.com/watch?v=OKnTUsQ6sb0&t=746s. For a more in-depth soundscape analysis of this song within its gameplay context, see my previous article, "Music as Temporal Disruption in *Assassin's Creed*," *The Soundtrack* 11, no. 1 (2020): 57–73.
78. Karen Cook, "Beyond (the) *Halo*," 199.
79. Karen Cook, "Medievalism and Emotions," 493.
80. A feature identified by Karen Cook in "'The Things I Do for Lust . . .'" as dating back centuries but particularly tied to "the eighteenth-century antiquarian conflation of numerous earlier types of bards and bardic activities" (29).
81. Karen Cook, "'The Things I Do for Lust . . .'" 27; Haines, *Music in Films on the Middle Ages*, 19, 71, and 108; and Randall Upton, "What Sounds Medieval?"
82. James Cook, "Playing with the Past in the Imagined Middle Ages: Music and Soundscape in Video Game," *Sounding Out!* (October 3, 2016), https://soundstudiesblog.com/2016/10/03/playing-with-the-past-in-the-imagined-middle-ages-music-and-soundscape-in-video-game/.
83. Ibid.
84. Haines, *Music in Films on the Middle Ages*, 71.
85. Randall Upton, "What Sounds Medieval?"
86. See, for example, "A Pleasant New Ballad, on Sir John Barleycorn: The Tune Is: Shall I Ly Beyond Thee" (London: [publisher not identified], 1670); and "A New Mad Tom of Bedlam, or, the Man in the Moon Drinks Claret: With Powder Beef Turnip and Caret, the Tune Is Grayes Inne Mask (London: [publisher not identified], 1690). Tom of Bedlam is also mentioned in Shakespeare's *King Lear* (1606).

87. Haines, *Music in Films on the Middle Ages*, 69.

88. Luke Plunkett, "*Assassin's Creed IV*'s Sea Shanties Are a Treasure," *Kotaku*, November 12, 2017, https://kotaku.com/assassins-creed-ivs-sea-shanties-are-a-treasure-1486865100.

89. Karen Cook, "'The Things I Do for Lust . . .'" particularly identifies features such as stepwise melody lines, modal music, and dance-like melodies as evoking the Anglo-Celtic folk style in addition to the folk instruments previously mentioned in this chapter (27).

90. *The Sims Medieval* (Electronic Arts, 2011), Microsoft Windows, MacOS X, iOS, and Windows Phone, music by John Debney.

91. James Cook cites the tambourine as particularly emblematic of medievalist tropes because of its on-screen association with medieval peasant dancing. James Cook, "Game Music and History," 357.

92. James Cook, "Playing with the Past."

93. *The Elder Scrolls V: Skyrim* (Bethesda Softworks, 2011), Microsoft Windows, PlayStation 3, Xbox 360, PlayStation 4, Xbox One, Nintendo Switch, PlayStation 5, and Xbox Series X/S, music by Jeremy Soule.

94. The scene can be viewed at https://www.youtube.com/watch?v=B9yR7S9whIM.

95. Carl Dahlhaus, *Studies in the Origin of Harmonic Tonality*, trans. Robert O. Gjerdingen (Princeton, NJ: Princeton University Press: 1991).

96. *Dragon Age: Inquisition* (Electronic Arts, 2014), Microsoft Windows, PlayStation 3, PlayStation 4, Xbox 360, and Xbox One, music by Trevor Morris.

97. Traxel, "Pseudo-Medieval Elements," 132.

98. Ibid.

99. The association of Celtic and Nordic cultures with medievalism is a frequent trope which "often signals a kind of rough-hewn but honest, democratic, community-based society, with elements of nostalgia, as well as a closeness to nature" (James Cook, "Game Music and History," 345).

100. Summers, *Understanding Video Game Music*, 93.

101. James Cook, "Game Music and History," 353.

102. *Samba de Amigo* (Sega, 1999), Arcade, Dreamcast, and Wii, music by Masaru Setsumaru.

103. Thomas B. Yee, "Racialized Fantasy: Authenticity, Appropriation, and Stereotype in *Super Mario Odyssey*," presentation at the North American Conference on Video Game Music, June 13, 2021.

104. Ibid.

105. A gameplay demonstration can be found at https://www.youtube.com/watch?v=RyU4g4MkGTU.

106. A gameplay demonstration can be found at https://www.youtube.com/watch?v=wYDFed_zvgg.

107. Buhler, *Theories of the Soundtrack*, 190.

108. James Cook, "Game Music and History," 352–353.

109. Karen Cook, "Music, History, and Progress," 177.

Three

Thematic Representation

While much of the previous two chapters focused on objective authenticity, examining how historical elements create new gameworlds that build from elements of the past but also recontextualize players' understanding of the past, the perception of authenticity by players ranges well beyond historical references. Player expectations and projections (what we identified in the introduction as constructive authenticity) are also key in how players understand authenticity. Does a game meet the player's expectations about its genre, mode of play, visuals, sounds, and more? How does a game form these expectations in the first place?

One common phenomenon in gaming is that of the game series or sequel, a means of reusing characters, gameworlds, and game mechanics to present new material to audiences in familiar formats. Constructive authenticity is invoked through this process, with the reuse and adaptation of earlier materials (particularly musical materials) serving as an easily accessible set of gameplay expectations but also as a mnemonic link to the narrative, tactile, and emotional experiences of previous gameplay. This is true of both narrative components within the game as well as other, more abstract elements; for example, the retro gaming genre (to be discussed in our next chapter) establishes player expectations based not necessarily on reused narrative elements, but rather by evoking the visual aesthetic, sonic landscape, and simple gameplay mechanics of earlier eras. Retro games might feel authentic because they conform to players' distorted memories of the past and as a result often invoke nostalgia in their players, an emotional response that corresponds to existential authenticity.

The next two chapters will focus on how constructive and existential authenticity are intertwined. In this chapter, I will examine the various ways in which music can reference previous gameplay experiences from a narrative perspective (be it in a previous game or within the same game), establishing player expectations. The use of musical themes is an important method by which players' intertextual experiences are invoked. While the concept of *Leitmotif* has been the main focus of most analyses in this area in the past, this chapter will present additional ways of using musical themes, textures, and instrumentation to create and reinforce narrative associations. The analyses of this chapter will discuss how pitch and rhythmic motives help to create narrative links but will also range beyond melody to focus on the role of instrumentation in evoking player memory, particularly between sequels. Lastly, this chapter will conclude with some analytical examples that explore the communicative potentials of thematic development.

THE VARIOUS ROLES OF THEMATIC REPRESENTATION

In our last chapter, we discussed how trope inflected our understanding of "Resurrection" from *Shadow of the Colossus*. Reverb, baroque-inspired counterpoint, and instrumentation such as organ and choir combined to create the trope of the sacred, deepening the player's emotional impact. But the song also fulfilled another musical role, with its reiteration throughout the game in conjunction with the characters Mono and Agro foreshadowing their resurrection. Rather than using musical topics as signifiers or combining them into referential tropes that provided signification through previous media experiences, the musical theme itself in this case acted as a signifier, referring to previous in-game experiences rather than previous out-of-game experiences.

This sort of thematic representation is often termed *Leitmotif*, the use of a recurring theme throughout a work of art (be it literature, music, or other formats such as the visual arts) such that the theme is associated with a particular character, location, situation, idea, or emotion. The leitmotivic melody in this sense becomes a signifier for the person, place, or thing, a signifier that can identify its reference without having to state it directly. This emphasis on metaphorical representation, the "saying a thing without saying it directly," has been a recurring theme in previous chapters, and this Leitmotif is no different, presenting yet another musical possibility for evoking concepts referentially. Unlike concepts such as trope, with Leitmotif the recognizability need not be from previous experience; rather, in our *Colossus* example the referentiality establishes itself in the course of its musical presentation.

As Matthew Bribitzer-Stull explains, Leitmotif functions not only as a musical theme, but also as an associative entity that accumulates meaning upon each repetition. As such, Leitmotifs can parallel dramatic developments within the narrative.[1] However, three questions arise about whether a melody might constitute a Leitmotif. First, what about situations where the melody does not accumulate meaning upon each repetition, but rather rehashes the *same* meaning repeatedly? Second, debate exists about whether Leitmotif requires a nuanced, developmental process or whether more superficial representations can be equally termed Leitmotif. Jessica Green, for example, argues that melodies should be kept simple in order to not lose their signifier function, stating that "once basic identifications are made with different themes, however, the Leitmotif can be modified or altered in order to reflect the changing status of the character, place, situation, or emotion."[2] This poses the question of whether Leitmotif requires thematic modification rather than simply thematic restatement. Finally, some scholars argue that a basic, unvaried (or too obvious) approach to Leitmotif weakens its signifying function, making it too superficial to allow for nuance. While scholars such as Rod Munday and James Buhler argue that its usage form doesn't matter in the end since the Leitmotif itself creates a *mythic* effect (to be discussed shortly)—the Leitmotif becomes a character of a sort within the media—it nevertheless raises the question of whether Leitmotif's usage must be subliminal rather than explicit to be most effective.

Bribitzer-Stull provides a solution to this problem by defining Leitmotif as a sub-category of what he terms "associative themes," and identifies other forms of associative themes such as *idées fixes*, musical symbolism, motto themes, and more that do not contain similar properties of development as Leitmotif.[3] While the focus of his text is specifically on Leitmotif, I would like to explore this concept of associative theme more broadly and how it manifests in video game music, and also explore how other musical properties not always associated with theme (such as distinct instrumentation) can also be recognized thematically. Let us recast our perspective by breaking this down into three distinct methods of musical theme as signification.

Thematic Representation through Character Association

From the early days of popular video gaming, sequels have been a popular means of bringing familiar content to its audiences. The first commercially successful video game, *Pong* (1972), for example, both repackaged its original *Pong* arcade machine as "Pong in a Barrel" and the more child-friendly *Puppy Pong*, but also released a two-player sequel within a year of the original release, taking advantage of *Pong*'s rampant popularity.[4] In both video games and film, sequels have particularly experienced a rise in popularity since the turn of the millennium, a rise that some argue has occurred both to balance audiences' desire for the familiar and the novel, and due to the sequels' potential for increased profitability.[5]

Many of the most popular video games of all time have built their popularity over several sequels, with series such as *Super Mario, Halo, The Legend of Zelda, Final Fantasy, Grand Theft Auto,* and *Pokémon* well known both within gaming communities and more widely within popular culture. These games have crossed over from video game to television and film and vice versa, creating wider media franchises with films such as the *Star Wars*, *Harry Potter*, and *Lord of the Rings* series inspiring new games and game series. While sound design between video game and film features significant differences in approach due to games' interactivity,[6] sound and music (in particular, recurring musical themes) have become a common way to create strong associations between various media franchise instantiations.

One common example in video games is the reuse of musical themes to represent specific characters within a game. This is particularly prominent when porting characters from one game to another, or to maintain player interest and continuity between game sequels, relying on the player's intertextual experiences and nostalgia for previous games in order to create attachment to these characters in new environments. Unlike Leitmotif, such thematic character representation is intentionally superficial and does not generally focus on thematic development. That said, despite this superficiality it is nonetheless the most emblematic way of connecting musical themes to narrative elements in video games. Take, for example, the character themes of the *Street Fighter* series. The musical themes that play on each level are determined by the simple choice of characters (so when Ryu faces off versus Guile, for example, either the Ryu or Guile theme will play). The theme repeats throughout the game when the associated character appears, with little to no modification, a repetition of theme that makes them instantly recognizable and also lodges them in the player's memory. When hearing the same melodic theme in later games within the series, players will recognize the melodic quotation from the original despite any changes in the song's orchestration and arrangement to match the current popular music aesthetic.

A comparison between two different arrangements of "Guile's Theme" from the *Street Fighter* series, *Street Fighter II* (1991; composed by Yoko Shimomura and Isao Abe) and *Street Fighter IV* (2008; rearranged by Hideyuki Fukasawa based on Shimomura's original theme), shows such a similarity.[7] In *Street Fighter II*'s version (henceforth "SF II"), the chorus consists of a one-bar pattern that is repeated, then a third bar that varies bar 1 by changing the rhythm of the neighbor-note pattern and concluding with an embellished arpeggiation of a major triad. The SF IV version retains a similar one-bar chorus motive, but instead of repeating the pitch upon repetition it transposes the motive down by third in its second bar. Additionally, its rhythms have been altered. It retains the rhythm of the opening neighbor-note motion (an eighth–sixteenth–tied sixteenth pattern), which allows players to recognize it as the same theme as SF II, but some of its rhythmic attacks are removed upon repetition. The verse of SF IV also simplifies the rhythm compared to that of SF II by removing repeated notes and lengthening others to produce a more regular eighth-note rhythm. The pre-chorus section of SF IV retains the original melody of SF II for its first two bars (not surprising since the original SF II melody included slower half-note and eighth-note durations, consistent with the overall rhythms already emphasized in SF IV's version) and only introduces a slight simplification of the rhythm and melody in its last four bars.

Thus there are enough common musical features between the music of SF II and SF IV for players to recognize the two themes as notably the same. Contour and interval are (for the most part) retained, and rhythmic changes still articulate the overall syncopation of the original SF II theme. The differences between the two themes manifest mainly in the accompaniment, which includes significantly different timbres. SF II uses digtially generated reproductions of bass, drums, synth, and trumpet, while SF IV uses a more realistic instrumental timbre, aiming for a 2000s hard rock aesthetic in the verse and chorus sections with electric bass, drum set, and electric guitar and contrasting "acoustic" piano in the pre-chorus section. Players still associate the melody with the Guile character despite these reorchestrations and rearrangements since the musical motive provides an aural marker of Guile-ness via its contour and syncopation. And therefore the music helps to fulfill player expectations (and thus constructive authenticity) in SF IV, where hearing the theme tells them Guile will once again appear, creating a clear connection between games in the series, and hints at the fact that the overall style of gameplay will remain the same, albeit with minor tweaks and improvements. As one YouTube commenter indicated in response to the song, "the legend of the man with two special moves continues."[8]

Interestingly, less change occurs when the characters are ported to other series. "Guile's Theme" in *Super Smash Bros. Ultimate* (2018), for example, retains the exact same melody and rhythm of *Street Fighter II*. Other than a modernization of its orchestration to use more realistic instrumental timbres, the arrangement of the song is the same as its 1991 rendition. *Super Smash Bros. Ultimate*, by retaining the 1991 musical arrangement, creates a musical link to the most famous game in the series and the iteration in which many of the series' character themes were first established. As such, it presents a more objectively authentic version of the song, smoothing over the discrepancy of experiencing these characters in a different media franchise than expected. The signification is simple—the presence of the song represents the presence of Guile and nothing more—but it is effective in evoking nostalgia in players of their earlier experiences with the *Street Fighter* series.

Thematic Representation through Gameworld Association

In chapter 1, I discussed how sound and music can establish gameworld through signifiers of time and place, narrative inflection, and game cueing. That discussion focused primarily on evoking gameworld within a single game, but gameworld themes can, much like character themes, create continuity between sequels by reminding players of their previous gameplay experiences. As previously discussed, themes may depict a particular geographical area or culture within the game by referencing the instrumentation, style, tropes, or stereotypes from real-world cultures, but this is not the only way that themes might depict gameworld. *Celeste* (2018), for example, changes its ostinato pattern, instrumentation (including shifts from acoustic timbres to retro digital sounds), and density of musical layering from one level of the game to the next, a sonic change that pairs with shifts to new geographic areas (new elevations, buildings to explore, or weather conditions) that the protagonist, Celeste, has attained, but which also reflects her evolving emotional state throughout the game.[9] The *contrast* of sound, rather than references to real-world cultures or geographies, establishes each level since the various themes are confined to the game itself and do not evoke external references, but the player nevertheless sonically understands a difference from one geographic area to the next that mirrors changes in both the gameplay mechanics and visual style (in particular, a different colour palette for each level).

Gameworld themes act to establish time and place in various ways and to provide detail to the world, but that is not their only function. As associative themes, gameworld themes create continuity for the player by telling them that they have returned to the same (or a related) location by returning to previously heard music. Take, for example, the "Temmie Village" and "Temmie Shop" themes from *Undertale*.[10] Although they use different instrumentation, both retain the same melody, indicating to the player that they are still within the borders of the Temmie region. However, as Zach Whalen describes in a different example (an analysis of *The Legend of Zelda: Ocarina of Time*),

> the three-dimensional construction of Link's environment often allows a player to choose whether or not to move toward the source of the "danger music," but the same broad structure of concluding a level or world with danger music holds true as Link encounters level bosses and the final enemy, Ganondorf. The application of this safety/danger binary in the fluid schematic of the three-dimensional space of Hyrule exhibits the complexity and richness of this fictional space.[11]

The two Temmie zones in *Undertale* depict their game function musically in the form of the safety/danger binary that Whalen cites. Timbrally, the song features a digitally morphed reproduction of the human voice that contains significant distortion. Paired with the sudden, unexpected transpositions of a semitone in the harmony and the repeated tonic-to-dominant bass (suggesting a polka style), the overall effect of the distortions and simple accompaniment structure is one of humour. The effect communicates to the player that this is a safe zone. This marker of place acts as a broad signifier rather than a specific one; while it communicates information to the player about the gameplay status in a specific area, the music is not necessarily prompting the player to specific actions and thus does not have a direct cueing function in the traditional sense. Iain Hart, on the other hand, presents examples where similar cues instead prompt the player to action, explaining that *Halo 3* "turns off the music after the player has spent too long in the one space to encourage them along and to minimize repetition," what he identifies as a *configurative sign*.[12] While the cue does not prompt the player directly, it does allude to potential future actions, setting up player expectations which, when met, establish constructive authenticity.

However, much like character themes, thematic associations to gameworlds (or areas within gameworlds) can also act as markers to evoke continuity between multiple games within the same series. An example from *The Legend of Zelda* series epitomizes this thematic representation of gameworld. In *Ocarina of Time* (1998), the first appearance of the Goron race within the series, the music of Goron City has distinct instrumentation of marimba, conga, cuica (a Brazilian friction drum), and a slide sound effect (possibly also derived from the cuica), a distinctly different sound than elsewhere in the game that clearly creates a sonic marker of that location.[13] These elements are reused in other games of the series, most literally *Majora's Mask* (2000), which reuses the exact same song and orchestration in its Goron-populated areas, but also in other varied forms. *Twilight Princess*'s (2006) "Death Mountain," for example, employs the same melody and accompaniment as the original song, and a similar instrumentation, but layers in a brass chorus and increases the activity of the percussion by adding a militaristic snare drum (see table 3.1 for a summary of the song's instruments); it also replaces conga with toms, two types of drum with similar timbres.[14] A more recent example, *Breath of the Wild*'s (2017) "Goron City—Day," retains the distinctive marimba, conga, and cuica accompaniment while layering in additional percussion instruments such as snare drum and cymbal, but slows down the overall tempo of the original song and adds an improvisatory trombone melody.[15] The music is nonetheless recognizably "the same," with the instrumentation,

Table 3.1. **Instruments associated with the "Goron City" theme variants throughout the *Legend of Zelda* series. *Source*: Table by the author.**

Game and Song	Melodic Instruments	Percussion Instruments
Ocarina of Time (1998) "Goron City" and *Majora's Mask* (2000) "Goron Village"	Marimba	Conga Cuica "Slide" sound effect
Oracle of Seasons/Ages (2001) "Goron Mountain/Rolling Ridge"	MIDI synth (4-track "retro" gaming waveform timbres: synth timbres)	
Twilight Princess (2006) "Death Mountain"	Marimba Trombone chorus	Snare drum Crash cymbal Toms (in lieu of conga)
Phantom Hourglass (2007) "Goron Island"	Marimba Synth organ/flute	Conga Cuica "Slide" sound effect
Spirit Tracks (2009) "Goron Village"	Marimba Synth organ/flute (Timbres in this game are more digital/MIDI than previous games in the series, and thus are digital interpretations of the listed instruments)	Conga Cuica "Slide" sound effect Snare drum
Skyward Sword (2011) "Eldin Volcano"	Marimba Orchestral string ensemble Brass ensemble (doubling strings)	Conga Shaker Glockenspiel-like sound effect
Breath of the Wild (2017) "Goron City—Day"	Marimba Synth organ/flute Trombone Tuba	Conga Steel drum Cuica Timpani Snare drum Crash cymbal
Breath of the Wild (2017) "Goron City—Night"	Marimba Synth organ Piano	Conga Steel drum Cuica

melody, and rhythm retaining enough similarity to that of *Ocarina of Time* for players to make the sonic association between the two.

In-game, this song is used to indicate specific geographic settings through its presence in areas populated by Gorons, but also in areas with similar geographies to *Ocarina of Time*'s Goron City, exploiting the players' intertextual knowledge of the series but also setting up expectations of certain geographic or gameplay properties. *Skyward Sword* (2011), for instance, uses a related song for its Eldin Volcano region, a volcanic, mountainous area with a similar rocky, cavern-embedded terrain to *Ocarina of Time*'s Death Mountain and Goron City. The region is not populated by Gorons, but a clear variation of the original song nevertheless is used, beginning with the same hand drum and marimba accompaniment (this time over brass) before introducing orchestral strings into the mix. The melody of "Eldin Volcano" is newly composed, and this variation to the song represents the same-but-different nature of the location, with a similar geography to Death Mountain but new non-player characters introduced. Most importantly, the association of this specific musical theme with these areas informs the gamer that their expectations from the Goron City and Death Mountain areas in previous games of the series will hold true in the new game; for example, dangerous rocks will fall down the mountain and lava will burn the player until they find the quest item that will protect them from fire and heat. The music thereby plays a functional role through this association.

Interestingly, while the instrumentation is derived from real-world musical genres, there is no attempt to link Death Mountain and Goron City to related real-world cultures. The conga, marimba, and cuica suggest a Latin American influence, but *Zelda*'s Gorons are a race of strongman rock creatures that if anything suggest sumo wrestlers (a similarity evoked directly in a mini-quest within *Twilight Princess*). The thematic association is made geographically but not meant to evoke the real world, making it distinct from the gameworld function of evoking time and place discussed in chapter 1.

Leitmotif

Because of its association with a specific place, one might ask why the Goron City theme does not constitute a Leitmotif. According to Bribitzer-Stull, three essential components are required for a theme to be considered a Leitmotif:

1. Leitmotifs are bifurcated in nature, comprising both a musical physiognomy and an emotional association. . . .
2. Leitmotifs are developmental in nature, evolving to reflect and create new musico-dramatic contexts. . . .
3. Leitmotifs contribute to and function within a larger musical structure.[16]

Interestingly, two components of this definition—emotional association and dramatic context—are important elements in Bribitzer-Stull's definition that point beyond the simple musical features of a theme. Furthermore, not all representative themes are Leitmotifs since development or evolution is also a critical element of his definition. "Goron City" might arguably constitute a Leitmotif in the *Zelda* series as a whole since the song evolves from one sequel to the next, but more likely it would not fit the definition for a number of key reasons. First, "Goron City" lacks a strong emotional association. While the player might feel a fondness for the Gorons because of their previous gameplay, this emotional association is more related to player nostalgia than the narrative context of the Gorons themselves. One could argue that the musical structure of "Goron City" might give an emotional subtext to the game, with its irregular accents and thinner textures musically depicting the often light-hearted interactions with the local Gorons; however, such an impression is more strongly depicted through the dialogue and physical gestures of the characters themselves rather than the music.

Second, "Goron City" lacks narrative nuance and development since the theme present in Goron City, Death Mountain, and similar locations does not vary, instead remaining static throughout the game. In some cases, the song is often duplicated exactly (or close to exactly) from one game to the next in the series, with similar rhythms, melodic structures, and instrumentation. The exact copy from *Ocarina of Time* to *Majora's Mask* is the most blatant instance of this, but other iterations such as that in *Oracle of Ages/Seasons* merely involve a superficial change of instrumentation and timbre. The usage of the song does not reflect much, if anything, in the way of narrative or dramatic development, but rather acts merely as local colour to establish a particular location or as a nostalgic evocation of previous games. The theme therefore, despite having an associative entity, does not accumulate meaning upon each repetition.

However, Leitmotifs do appear in other contexts in gaming. Perhaps inspired by its usage in opera and film, Leitmotif in video games is driven by a narrative approach to music and aims to create emotional responses in its audience that heighten and intensify the dramatic experience.[17] Unlike in film though, video games add a layer of interactivity to audiences' interpretations of Leitmotif, with the possibility of impacting how players make choices within the game. Leitmotif can also fulfill several roles. While many scholars emphasize Leitmotif's role in creating unity throughout an artwork and its derivation from the compositional practices of Richard Wagner,[18] from a media perspective its more important function is to communicate unspoken thoughts or subliminal messages in media. As Justin London describes, "musical leitmotifs . . . do more than simply designate; they also contain an expressive content that is entwined with its musical structure.... They couple a capacity to refer with a sense of emotional expression."[19] Some common usages for Leitmotif in video games include narrative foreshadowing (such as the introduction of a character before they arrive on-screen) or to allude to the merger of two previously-distinct plot elements.

The revelation of a character's hidden identity is a media trope in video games that frequently employs this strategy. An oft-cited example of this phenomenon occurs in *The Legend of Zelda: Skyward Sword*. The song "Ballad of the Goddess" is a melodic and rhythmic retrograde of "Zelda's Lullaby," heard in previous games in the series such as *A Link to the Past* and *Ocarina of Time*. However, this relationship is not immediately apparent upon first hearing given the difficulties of aurally perceiving retrograde relationships. As Fred Lerdahl explains, "retrograde is troublesome to grasp by ear even if intervallic direction is preserved, retrograde inversion still so more."[20] On the other hand, the reversal of the melody retains common elements such as tempo and its triple meter (producing a rocking lullaby feel), which evokes a sense that the songs are similar even if the player does not detect the thematic association between the two. Once the player realizes the relationship between the two melodies, they become aware that the game is foreshadowing Zelda's identity as The Goddess; narratively, the retrograde transformation of the Leitmotif here obscures the melodic identity in much the same way that Zelda's knowledge of her own identity is obscured. The motivic transformation parallels the narrative, and the song acquires new meaning as the player progresses in the game.

However, an important question here is whether the player needs to be aware of the Leitmotif's connection to the previous theme in order for it to function narratively. What if the player does not recognize the melody as the retrograde of the original, but instead hears it as a distinctly new theme? London suggests that Leitmotifs should be musically distinctive, relatively stable, and recognizable, which may not be the case in this instance.[21] Furthermore, what about players who have not played a previous *Zelda* game and are thus unfamiliar with "Zelda's Lullaby"? If Leitmotif's strength is in its particular connection to melody, motive, and the recognizability of this motivic content, this function is absent if its audience is missing its referent, and the building of player expectations that defines constructive authenticity is no longer present. This poses the question of whether Leitmotifs can exist across an extended media franchise or series or whether they are limited to a single iteration within a series (certainly if we are modelling our understanding of Leitmotif after Wagner and his *Ring Cycle*, then the potential for leitmotivic thematic association across a series exists).

On the other hand, some scholars critique the too-obvious use of Leitmotif. Munday summarizes these critiques, observing that

> cinematic music can be criticized when it is used in video games because it acts merely as a kind of quotation of the epic music found in cinema, connoting all the epic associations of the aesthetic of cinematic realism, but without actually performing any of its functions . . . reducing leitmotif to the lowly status of a looping underscore.[22]

However, in the same chapter he articulates that video games function quite differently from film, and thus these critiques may not be applicable since video games use less subtle "signpost" Leitmotifs to indicate safety versus danger and other important gameplay cues that structure player expectations.[23] And other scholars such as Buhler recognize that "the demythifying impulse of film music leads not away from myth but back towards it. This is the riddle of the leitmotif, which entwines myth and signification in a knot almost impossible to solve."[24] Thus the simplicity of the motivic presentation is irrelevant to its ability to evoke a mythic status. More recent developments in video game music over the last fifteen years have used Leitmotif in contexts beyond mere signposts (and indeed, this is the reason why I am distinguishing between Leitmotif and other forms of thematic representation in this chapter).

Leitmotifs can, however, employ this mythic effect to pull players into a gameworld and help to establish legitimacy (and thus authenticity) within their larger cinematic universe.[25] For example, in *Star Wars Jedi: Fallen Order* (2019),[26] fragments of the well-known Force Theme (established in the original trilogy, 1977–1983, and as identified by Frank Lehman, used in all nine of the trilogy films as well as *Rogue One* and *The Mandalorian*[27]) are used at key moments of the narrative to both foreshadow future events and link to the cinematic universe's larger mythology.[28] One such example occurs when the protagonist, Cal, discovers a hidden vault. The first seven notes of the Force Theme, in its characteristic French horn instrumentation, play as the door to the vault opens and immediately repeat transposed up one semitone (just in case the player didn't catch it the first time!).[29] The brief evocation of this theme indicates to the player, without stating it directly via dialogue, that the vault has a connection to the Jedi and the Force and also serves to remind them of the game's mythos.

As Claudia Gorbman describes, a common strategy for presenting Leitmotifs employs the opening measures of themes rather than their full forms.[30] That is true of our example: *Fallen Order*'s soundtrack avoids using the full Force Theme quotation from the original trilogy, instead moving on to newly composed music. As one fan describes, "it feels like it was actually produced by [original *Star Wars* composer] John Williams. It captures the same feel of the Lucas' star wars [sic] in a way I haven't seen a game do before, and not by putting the old OST's on repeat."[31] The fragmentary usage mitigates the too-obviousness of the signpost effect since there is just enough of the theme to recognize before new musical content is introduced, and this new musical content continues to evoke the mysticism trope through musical topics such as chime timbres and high strings in the orchestration rather than through more extensive quotation. Furthermore, the mythic effect is intentional here. Yes, the Leitmotif is extremely obvious, but players are expecting such referential melodies as part of the gameplay experience and actually complain when they are not present. For instance, in the same fan discussion just cited, another player complained that "I get it, yes it sounds like Williams, but the fact that I'm like five hrs in & haven't once heard the FORCE THEME or any other stuff that truly gives starwars half it's [sic] magic is really disappointing me." *Star Wars* fans are expecting a reference to the cinematic universe's in-world mythology, meeting their expectations of connections to the cinematic universe to build constructive authenticity, and John Williams's establishment of Leitmotif in connection with that universe from its cinematic beginnings becomes part of the authentic experience of that series.[32] As Buhler argues, "the music seems to intuit connections that are beyond immediate rational comprehension. . . . Music is therefore fundamental to the representation of the Force."[33]

Not all usages of the Force Theme are as effective as this one. In other games, the theme is used to try to establish the game as part of the larger *Star Wars* cinematic universe, but not always in narrative contexts that make sense. Take, for example, the use of the Force Theme motive in *Star Wars: Knights of the Old Republic* (2003).[34] In one scene, the protagonist is knocked unconscious, and upon waking the Force theme plays while they meet their new companion in a safe zone.[35] There is no obvious narrative connection to the Force at this point, the connection to Jedi is not explained until the new companion details the protagonist's new mission objectives, and even then the mention of Jedi or other referents for the Leitmotif, such as fate and destiny, is brief at best. The theme feels a bit too obvious here, trying too hard to suggest the *Star Wars* universe without a clear narrative connection to the Force. The border between creating an authentic experience within the cinematic universe and pulling the player out of their mental conceptualization of the magic circle is a fine line to walk.

MORE ADVANCED APPLICATIONS OF THEMATIC REPRESENTATION

More nuanced applications of Leitmotif tend to succeed much better in retaining gameworld authenticity than more obvious signpost effect

108   *Chapter Three*

usages, as do those that undertake several narrative functions or thematic representation methods simultaneously. Notable is the fact that the three means of thematic representation through music discussed in this chapter do not exist discretely, but rather overlap significantly with tropes/media tropes and player intertextualities. Game series in particular make use of Leitmotifs, necessitating the knowledge of the previous game. The thematic representation functions also overlap significantly, with a single melody manifesting, for example, as both a Leitmotif and a character theme. Such a multifunctional approach itself colours players' understanding of narrative, and often evokes nostalgia in players for their previous gameplay experiences, engaging with emotion in ways that suggest existential authenticity. Let us take a look at one such nuanced example.

Thematic Foreshadowing via Leitmotif

One of the most famous first-person shooter series, *Halo*, the flagship game of the Xbox, has consistently used thematic representation through character themes, Leitmotifs, and more, reflecting a growing narrative construction over its twenty-year life span. The game is perhaps best recognized via *Halo*'s main theme, given in figure 3.1 (the boxes indicate subsets used for motivic development that will be discussed shortly).[36] This melody, a chant for unison male voice set against orchestral accompaniment, uses the plainchant trope to suggest the ancient culture and ring-shaped worlds religiously worshipped by the game's main antagonist.[37] As Karen Cook describes, "plainchant, for [the game's composer] O'Donnell, is thus recognizably imbued with and able to impart the qualities of oldness, mystery, fantasy, sacredness, even the alien, not only for him but also for his expected listeners."[38]

Players' familiarity with this main theme was exploited by co-composer Martin O'Donnell, who employed the theme developmentally

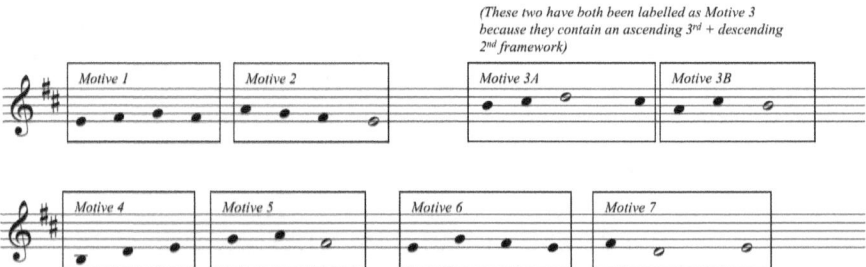

Figure 3.1. Chant melody from *Halo*'s main theme, notated in unmetered rhythm with breakdown into sub-motives. *Source*: Composed by Martin O'Donnell. Reduction and analysis by the author. Original music copyright Xbox Game Studios.

in the music of the game's sequels. O'Donnell, in a 2007 interview with *Wired* magazine, discusses the composition of *Halo 3*'s "Keep What You Steal," describing how he explicitly reused musical content from previous games, including the original theme as well as a related chord progression from *Halo 2*. He states:

> O'DONNELL: I don't know why, it's funny because when I did that [chord progression] I thought: is that over the top? Is that going to be, like, too emotional? Are people just going to start laughing because it's just, like, it's just this dude coming out of the smoke, he's got a spacesuit on—I mean, come on!
>
> INTERVIEWER: It's not just some dude.
>
> O'DONNELL: It's you! You're back![39]

O'Donnell's comments make it clear he is well-aware of the signpost effect here, using an obvious quotation from the previous game to create continuity between the two series. The interviewer's response articulates players' identification with the spacesuit-clad main character, Master Chief, and how his return creates a sense of heroic mythos as well as an emotion of nostalgia for the main character, building existential authenticity.

O'Donnell's treatment of theme, however, is also rich in motivic development. Let us take a look at "Luck," the music that plays over the opening cut scene of *Halo 3* (a reduction of bars 17–57 are given in figure 3.2).[40] The cut scene begins with an image of stars accompanied by a voice-over from Cortana, the protagonist's digital companion. This voice-over alludes to several plot points from previous games in the series, such as Cortana's selection of the protagonist (named "Master Chief" throughout the series), his heroic actions and super-human abilities, and their close friendship, but for those who have played *Halo 2* the presence of Cortana's disembodied voice also acts as an unspoken reminder that she was lost in an act of self-sacrifice at the end of that game. The voice-over continues as the cut scene shows a comet in the distance coming ever closer, soon revealed to be Master Chief crashing to Earth. When Master Chief shakes off his forceful landing and revives, he is asked by Johnson (his sergeant, a familiar character from previous games in the series), "Where is she, Chief? Where is Cortana?-" Johnson's question begins at bar 43 of figure 3.2 and continues over the next four bars, explicitly linking the theme heard in the flutes at this location with the question of Cortana's fate.

This flute motive is derived from *Halo*'s main theme, creating a leitmotivic link to the previous games. On figure 3.1 I have parsed *Halo*'s main theme into eight sub-motives of three to four pitches each to give a point for comparison with the new motivic material, but also because

110 Chapter Three

O'Donnell's compositional style suggests a process of fragmentation (breaking motives down into smaller forms), liquidation (the "systematic elimination of characteristic motives," as defined by William Caplin), and variation.[41] For example, in the main theme, the motives I have labelled as 3A and 3B are closely related: both contain a similar intervallic contour of an ascending third plus descending second (pitches 1, 3, and 4 in Motive 3A and 4, 5, and 6 in Motive 3B), but 3A contains an additional pitch, C♯, to fill in the ascending third interval. We could thus consider 3B to be a simplification of 3A to its essential framework. This two-fold repetition of Motive 3 is, in fact, the exact motive used for Cortana's flute theme in bars 44–48 of "Luck," thereby recalling the events of *Halo 1* and *2* but also accumulating new meaning as Cortana's theme.

Other motivic content in "Luck" is also derived from *Halo*'s main theme. For example, the melody starting in bar 18 begins with the same three pitches as Motive 1 of the main theme (E, F♯, G); rather than continuing on to Motive 2, it instead begins a chain of our Motive 3 (a third + second interval string) just discussed in bars 19–22, followed by two notes to conclude the phrase that are reminiscent of the ending of the

Figure 3.2.   Motivic analysis, bars 17–56, "Luck," *Halo 3*. *Source*: Composed by Martin O'Donnell. Reduction and analysis by the author. Original music copyright Xbox Game Studios.

main theme's concluding pitches. Bars 25–26 follow the same descending contour as Motive 2, and bars 25–27 state Motive 1 again.

More significantly, though, Motive 3 serves as the basis of significant motivic development within "Luck" itself. Motive 3 is developed slightly differently in three melodic statements: bars 19–21 (the chain of Motive 3s described above), bars 33–37 (a new theme introduced just before Johnson's question about Cortana's whereabouts which inverts the contour of the original Motive 3), and our Cortana flute theme of bars 44–48. In figure 3.3, I have aligned these themes, transposing them to the same octave for the sake of comparison. All three share D, C♯, B within their first four notes, and the latter two motives also conclude with a common A, B. But each is a slightly different iteration, adding pitches to vary this underlying melodic framework. In "Luck," this process of motivic development culminates in bars 51–56 where we hear a significant change to a slower tempo and a shift in orchestration to emphasize piano. At this moment, we hear two final developments of Motive 3 that again share the same framework. This time, though, the new theme appears to share the added C♯ from bars 44–48 and the added initial E from bars 51–53, and also adds triplet ornamentation.

We thus have a process of developing variation, building on a previous variant rather than the original source motive. As Ethan Haimo defines it, developing variation occurs when:

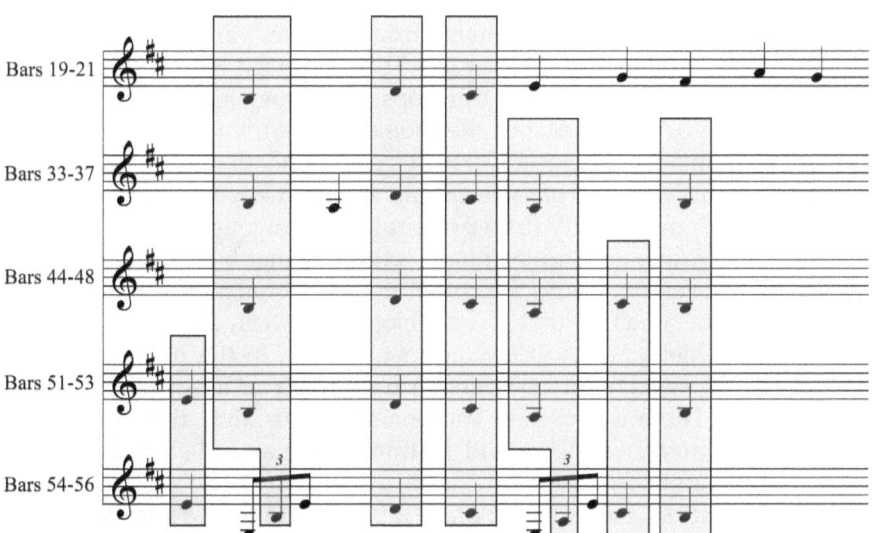

Figure 3.3. Melodic development derived from Motive 3 in "Luck." Boxes highlight pitches in common between the five theme variants. *Source*: Figure by the author.

1) an initial motivic figure is stated; 2) this is followed by another figure that retains enough of the features of the first to be recognised, at least in some dimensions, as a restatement of that figure, but with significant changes in other dimensions; 3) those changes effectively create new musical configurations which can then be subjected to further development by restatement combined with variation.[42]

True developing variation can be distinguished from purely local varied repetitions that have no developmental consequences. Developing variation offers the possibility of forward motion, permitting the creation of new or contrasting (but still related) ideas, while local variation affects only the passage in questions.[43]

Cortana's theme is a true instance of developing variation in that it is clearly no longer the original *Halo* main theme despite having a clear derivation from this material. It lacks Motives 1, 2, 4, 5, 6, and 7 of the main theme and is instead only constructed from Motive 3, and new contexts for this motive provide additional variation. The new melodic idea thus created becomes clearly associated with Cortana through narrative context and reappears later in the game in association with her reappearance in a decidedly leitmotivic way.

Later in *Halo 3*, when Cortana is finally found, we hear similar thematic material and instrumentation in a new song titled "Keep What You Steal," the opening eight bars of which are analyzed in figure 3.4. The same pitch classes and intervallic pattern of the Cortana flute theme from bars 44–48 of "Luck" (B, D, C♯, A, C♯, B) can be heard in the piano's highest register (boxed) of "Keep What You Steal"; however, this new instance incorporates musical elements from the later variations of "Luck," including the initial opening E of bar 51 and the triplet motives of bars 54–56. The Cortana theme thus develops over time, employing thematic representation of character, but the theme also employs distinct properties of Leitmotif, representing Cortana's earlier relationship with Master Chief through the reuse of *Halo*'s main theme. However, it transforms the theme both motivically (to represent their evolving relationship) and instrumentally, using a similar piano setting to that heard in bars 51–58 of "Luck." This also employs the common "emotional piano" media trope to represent the affection that has developed between the two characters and Master Chief's regret at having lost Cortana in the previous game, connecting the two characters together narratively as a central plot point of the series. There is a certain component of nostalgia that is inherent to the use of character theme and Leitmotif here as well, which we will explore in the next chapter.

Instrumentation, then, clearly plays a role in the theme's recognizability, and this is not the only point in the music where it helps the player

Thematic Representation 113

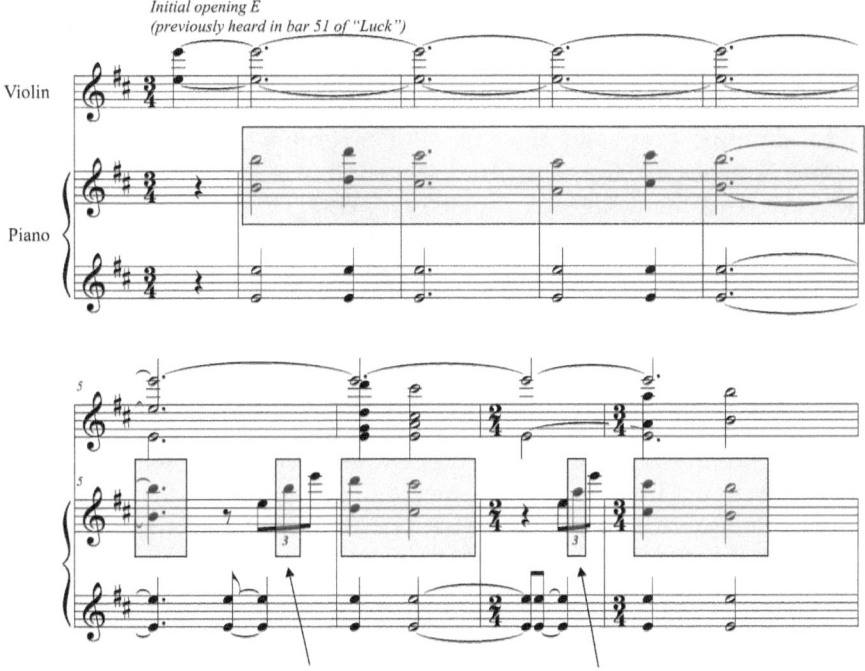

Figure 3.4. "Keep What You Steal" excerpt, *Halo 3*. Boxes indicate material derived from previous melodic content in "Luck." *Source*: Composed by Martin O'Donnell. Reduction and analysis by the author. Original music copyright Xbox Game Studios.

to make both musical and narrative associations. In "Luck," for example, I examined how bar 18 recalled the opening three notes of *Halo*'s main theme. One of the distinctive properties of the main theme from the beginning of the series was its low-register (digital) vocal timbre, and in bar 18 the low register of the double bass evokes a similar effect. Despite being performed by a live choir and orchestra, "Luck" makes prominent use of an untexted choir whose "ah" timbre evokes the original main theme (this can be best heard in the example starting on the pickup to bar 5). The instrumentation and timbres are thus a component of the theme's recognizable musical identity, recontextualized anew. As Thomas Grey describes, "Leitmotif, then, is not just a musical labeling of people and things (or the verbal labeling of motives); it is also a matter of musical memory, of recalling things dimly remembered and seeing what sense we can make of them in a new context."[44] Put another way, players' musical memories become a fount of knowledge to draw upon when setting expectations for the new context, leading to constructive authenticity.

ASSOCIATIVE THEMES WITHOUT CHARACTER OR GAMEWORLD?

It is clear that a recognizable musical identity and its repetition is required to create the three types of thematic association (character, gameworld, and Leitmotif) examined thus far. But are other types of associative themes possible? Do they still build authenticity in players? Let us examine a recurring theme in *Sonic the Hedgehog* (1991) that associates not with character or gameworld, but with game function: the main title theme music.[45] Although much of Sonic's music is geared towards reflecting the fast, forward-moving momentum of the game's main character,[46] the music of the main title theme (henceforth "Title Theme") does not necessarily follow this momentum, instead recontextualizing the theme with increasing amounts of irregularity as the game progresses based on new gameplay contexts and functions. Variations of the theme recur as the Title Theme, the Continue menu, and the Game Over screen, all sections that do not involve active gameplay with control of the avatar, but rather communicate game state to the player. This process helps to create anticipation and expectation, supporting both constructive and existential authenticity.

Particular elements of musical variation contribute explicitly to the player's understanding of this recurring theme and how it communicates new information through musical recontextualization. Let us start by examining the Title Theme (figure 3.5), which will give us a point of comparison for later iterations. The theme is in a regular meter, 4/4, supported by consistently spaced off-beat percussion attacks (with the exception of a shift to an on-beat note at the moment of cadential arrival in bar 3). Motivically, it begins with two statements of a basic idea featuring similar rhythms and contours, a repetition that helps to reinforce the motive to the player and also establishes this as a recurring theme (and thus the beginning of both the music and the game).[47] Contrasting material occurs in bar 3, with a different rhythm (omitting the tie, for example, from beats 3–4 and including syncopations in the countermelody and bass parts) and significantly different contours in all voices compared to the previous bar. Notably, beats 3–4 in bar 3 create a strong rhythmic arrival, with the syncopation on beat 3 creating tension that resolves with the on-beat arrival of beat 4, and the ascending semitone motion further emphasizing this tension-and-resolution movement. Beat 4 thus constitutes a clear cadence despite its lack of traditional cadential harmony, with the bass octave leap further reinforcing its sense of finality.

As David Huron describes in *Sweet Anticipation*, many Western-based musics have a preference for binary divisions.[48] Huron first discusses this in terms of beat groupings and subdivisions, but consequently expands the idea to include larger-scale formal structures such as phrase. The

Figure 3.5. Title Theme with phrase segment analysis, *Sonic the Hedgehog*. **Source: Composed by Masato Nakamura. Transcription and analysis by the author. Original music copyright Sega.**

consequence of this preference for binary divisions is that we generally expect a regular phrase length to be a binary-based length of 2, 4, 8, or 16 measures long. The final bar of the Title Theme does not include any melodic content, instead only prolonging its bass note. This acts to create a mild irregularity of phrase length. In a standard phrase we would normally expect the tonic arrival of the cadence to occur on the downbeat of the fourth bar due to this preference for binary divisions, but here it occurs one beat early in bar 3, suggesting a compression of the expected phrase length.

A modified version of the same theme is heard in two other menus within the game. When the player dies, a "Game Over" screen appears, followed by a "Continue?" screen that gives them the choice of whether to keep playing or quit the game. Both screens use music based off the Title Theme, framing player expectations based on earlier gameplay, but notably introduce features that disrupt the original theme's stability in ways that suggest each screen's game function, building tension in the player.

Sonic's Game Over Theme (figure 3.6) begins with the same basic idea and repetition originally heard in the Title Theme, but at a significantly slower tempo. However, the third bar of the Game Over Theme changes its contrasting idea. Rather than the original ascending leaps from the Title Theme, we now hear a new descending chromatic "lament" motive. While the longer duration of the Game Over Theme's final note does suggest a rhythmic arrival and cadence, this moment is also less stable in several ways. It features an exaggerated vibrato, is harmonically unstable

Figure 3.6. Game Over music with phrase segment analysis, *Sonic the Hedgehog*. Source: Transcription and analysis by the author. Composed by Masato Nakamura. Original music copyright Sega.

due to the dissonant major second between the bass's F♯ and the melody's G♯, and the percussion part drops out completely. While Leitmotif is typically defined as a recurring theme associated with a particular person, idea, or dramatic situation, this theme, despite not being associated with a character or element of the narrative, does present thematic association and repetition and recontextualizes its original musical setting upon successive iterations, key components of Leitmotif. Such changes structure player expectation and reaction. In the Game Over Theme, for instance, the transformations of the music compared to the Title Theme produce a sense of ending, fitting for its Game Over function, but also a sense of disappointment for the player that the gameplay is coming to a close. The lack of harmonic and melodic resolution, as well as a move away from the A-major tonality through chromatic motion in the new contrasting idea, suggest a lack of fulfillment. The slowing of both the overall tempo and the percussion dissipates musical energy, as does the omission of percussion in the final motive. In my own gameplay with friends and students, players and others in the room typically burst out in laughter, not at the moment of failure a second or two earlier, but instead at the moment the Game Over Theme begins, suggesting that the audience may in fact be reacting to the musical dissipation and distortion from expected norms of stable tempo and pitch. This distortion is not read as *inauthentic* in my view, but rather as humorous commentary still stylistically consistent with the rest of the game.

The Continue Theme (figure 3.7) similarly varies the original Title Theme in ways that communicate new information to the player about the game state. Like the previous two themes, the Continue Theme begins with the same basic idea in its first bar, but unlike the previous two themes the Continue Theme repeats the basic idea three times, each time modulating it to a new key (A major in bar 1, B♭ major in the melody and

Figure 3.7. Continue music with phrase segment analysis, *Sonic the Hedgehog*. *Source*: Transcription and analysis by the author. Composed by Masato Nakamura. Original music copyright Sega.

bass in bar 2, and B major in the melody and bass in bar 3). However, as the countermelody remains in A major throughout, this results in an increasing amount of dissonance each bar; by the time we reach bar 4 three different keys occur simultaneously to create a jarring dissonance that the player can only resolve by making a choice to continue or quit the game.

The final bar of the Continue Theme once again presents a new idea, but unlike the previous two themes it is not especially contrasting to the basic idea, and it does not include features that suggest a cadential arrival. While the melody in the final bar reverses the melody's melodic contour, it retains a very similar rhythm to the basic idea; the countermelody and bass parts retain the rhythm and contour of their basic idea. We do not hear a prolonged note to end the Continue Theme, and it lasts a full four bars rather than the "three bars plus a held note" pattern of the Title Theme and Game Over Theme. In spite of its length, which appears more regular at four bars, the Continue Theme sounds much more *irregular* than the Title Theme. Why is this the case? One element is likely the syncopation that ends each repeat of the basic idea here. While the syncopations in bars 1–3 resolve to a stable downbeat in the countermelody and bass at the start of the next measure, after measure 4 no such strong-beat attack is heard. Second, in the Title Theme, a displacement of the percussion's off-beat rhythm created an on-beat attack at the cadential arrival, but no such displacement occurs here in the Continue Theme to suggest an ending. Third, there is no shift to a contrasting idea or agogic accent in the melody to suggest an ending. Once again, these modifications suggest a process of development and variation typical of Leitmotif, communicating narratively about the game (or game state, in this case). The music's four-fold repetition and lack of tonal or rhythmic stability suggests a desire for continuation, which is appropriate given that the "Continue?" menu itself is prompting the player to begin the game anew. Once again, player expectations are disrupted, but in a manner that is still musically cohesive with the player's previous experience.

Huron posits that the listener will extrapolate predictability or regularity through familiar patterns, which in the case of our three examples is established through the recurring one-bar basic idea. He also argues that irregularities can be perceived as normal depending on context, that "listeners have little difficulty predicting nonperiodic event patterns, as long as the patterns are familiar."[49] In these examples, players are expecting a repetition of the basic idea and the introduction of a contrasting idea but are still able to recognize the themes as related despite their modifications. Even though each of the three themes features musical irregularities, all three feature adjustments to their phrase structure, tonality, texture, and tempo that make them optimally suited for each of their respective menu roles throughout the game. Musical irregularity and instability, rather than sounding unnatural, contribute to the cohesiveness of the game by augmenting the player's sense of anticipation, continuation, and conclusion as the game progresses, but also by structuring expectations. Despite experiencing each menu as an interruption to their game continuity, the menu musics feel *authentic* within the established framework of the game because their variation processes mirror the changes happening in their game state, conforming to the gamer expectations established within constructive authenticity. Thus musical authenticity can be established not only in narrative contexts, but also in modes connected to gameplay function.

THEMATIC REPRESENTATION: INTEGRATING THE WHOLE

While I started this chapter by breaking down thematic representation into the three categories, I have shown through subsequent examples that associative themes often fulfill more than one of these functions. Consequently, it may be more logical to think of them as three facets of the same phenomenon rather than three distinct processes. Similarly, Bribitzer-Stull describes Leitmotif as a form of conceptual integration, semantic mappings that "by making manifest the emotions that surrounded a certain dramatic moment of the past . . . [allowing] the composer to inject this experience into a new musico-dramatic context. As such, leitmotivic association sits at the intersection of emotion, memory, and meaning."[50] This "intersection of emotion, memory, and meaning" has a tantalizing similarity to our three components of authenticity, with emotion central to existential authenticity, memory and meaning essential in producing constructive authenticity, and memory further extending into our mediation of objective authenticity.

The more complex analyses in this chapter also suggest that trope, media trope, and thematic representation are much more closely linked than at first glance. Bribitzer-Stull fully acknowledges this in his discussion of Leitmotif, stating that it is often difficult to distinguish the processes that produce associative themes with those that produce references from pre-existing cultural media tropes.[51] Indeed, both processes boil down to players' intertextual knowledge structuring expectation. In many (but again, not all) of these cases developing variation seems to play a strong role in creating recognizable motivic relationships. *Halo 3*'s thematic breakdown and reconstruction, for instance, mirrors how Arnold Schoenberg describes developing variation in the music of Brahms: producing "new [melodic] skeletons by changing the viewpoint as to which are the main features and which are subordinate."[52] This manner of producing new material from old promotes thematic and narrative association—much like the narratives of video games themselves (and their sequels!).

While thematic representations of character and gameworld may go one level further in mediation, pulling on the player's emotions towards the character or gameworld as a whole rather than towards one exact narrative moment, they nevertheless bring past experiences to the present and in many cases create new associations in the process, evoking both constructive authenticity, through its establishment of expectations, and existential authenticity, through the manipulation of player emotion. That said, thematic representations in video games sometimes call for a simpler mode of evoking the past, simply restating previous themes rather than following extensive processes of variation. Is this a less effective means of thematic representation? In my view, no. The signpost effect in many gaming contexts simply acts as a quick, efficient tool to pull previous gameworld knowledge into the player's expected narrative and gameplay parameters. Authenticity is much more about conforming to player expectations (even if that expectation involves breaking expectations in stylistically consistent ways) than about creating nuance.

NOTES

1. Matthew Bribitzer-Stull, *Understanding the Leitmotif: from Wagner to Hollywood Film Music* (Cambridge: Cambridge University Press, 2015), 10.

2. Jessica Green, "Understanding the Score: Film Music Communicating to and Influencing the Audience," *The Journal of Aesthetic Education* 44, no. 4 (2010): 88.

3. Bribitzer-Stull, *Understanding the Leitmotif*, 10.

4. The International Arcade Museum, "Pong Doubles," in *Museum of the Game*, accessed August 30, 2021, https://www.arcade-museum.com/game_detail.php?game_id=9075.

5. Michael Pokorny, Peter Miskell, and John Sedgwick, "Managing Uncertainty in Creative Industries: Film Sequels and Hollywood's Profitability, 1988–2015," *Competition & Change* 23, no. 1 (2019): 24–26.

6. See, for example, Mundhenke, "Musical Transformations from Game to Film in *Silent Hill*," which examines the different approaches to sound setting in the *Silent Hill* franchise.

7. *Street Fighter II: The World Warrior* (Capcom, 1991), Arcade, Super Nintendo Entertainment System, Amiga, Atari ST, Commodore 64, ZX Spectrum, and PC (DOS), music by Yoko Shimomura and Isao Abe; *Street Fighter IV* (Capcom, 2008), Arcade, PlayStation 3, Xbox 360, Microsoft Windows, iOS, Android, PlayStation 4, and Xbox One, music by Hideyuki Fukasawa. The official soundtracks are available on Spotify at https://open.spotify.com/album/4jkBDRFtxJ0lKWYcqRSrGZ?si=oqSWyBynTXqsfKkAEJ2LWw and https://open.spotify.com/album/1DRWIHpmCo3H13NddOTAZu?si=DPY_CjbDTC-8PK7CIN0jtw.

8. "Super *Street Fighter IV*—Theme of Guile," accessed March 6, 2022, https://youtu.be/Q6XM3fDUxLg.

9. *Celeste* (Extremely OK Games, 2018), Linux, MacOS, Microsoft Windows, Nintendo Switch, PlayStation 4, Xbox One, and Stadia, music by Lena Raine.

10. *Undertale* (Toby Fox and 8-4, 2015), Microsoft Windows, MacOS X, Linux, PlayStation 4, PlayStation Vita, Nintendo Switch, and Xbox One, music by Toby Fox. The audio is available via https://open.spotify.com/track/7ym2c9z3JgkXVEQDk0sKn8 and https://open.spotify.com/track/3SVrOYdMaqeW56eHfw7ScX?si=b0f5adb5ada34013.

11. Whalen, "Play Along."

12. Hart, "Semiotics in Game Music," 225.

13. *The Legend of Zelda: Ocarina of Time* (Nintendo, 1998), Nintendo 64 and GameCube, music by Koji Kondo.

14. *The Legend of Zelda: Majora's Mask* (Nintendo, 2000), Nintendo 64, music by Koji Kondo and *The Legend of Zelda: Twilight Princess* (Nintendo, 2006), Wii and GameCube, music by Toru Minegishi and Asuka Ohta.

15. *The Legend of Zelda: Breath of the Wild* (Nintendo, 2017), Nintendo Switch, music by Manaka Kataoka, Yasuaki Iwata, and Hajime Wakai.

16. Bribitzer-Stull, *Understanding the Leitmotif*, 10.

17. Rob Munday, "Music in Video Games," in *Music, Sound and Multimedia: From the Live to the Virtual*, ed. Jamie Sexton (Edinburgh: Edinburgh University Press, 2007), 58. I will not go in depth into Leitmotif's history here as it is outside the scope of this focus on video games; a more thorough study can be found in Bribitzer-Stull, *Understanding the Leitmotif*, 20–30.

18. Justin London, "Leitmotifs and Musical Reference in the Classical Film Score," in *Music and Cinema*, eds. James Buhler, Caryl Flinn, and David Neumeyer (Hanover, NH: University Press of New England, 2000), 85.

19. Ibid., 89.

20. Fred Lerdahl, *Tonal Pitch Space* (Oxford: Oxford University Press, 2001), 378.

21. London, "Leitmotifs," 88.

22. Munday, "Music in Video Games," 60.
23. Ibid., 62.
24. James Buhler, "*Star Wars*, Music, and Myth," in *Music and Cinema*, eds. James Buhler, Caryl Flinn, and David Neumeyer (Hanover, NH: University Press of New England, 2000), 43.
25. "Cinematic universe" is being used here according to its popular definition. For example, the "Marvel cinematic universe" is the fictional shared universe established by the Marvel movies, as suggested by its name, but which also includes video games and comic book plots set in the same grouping of canonic storytelling/media franchise.
26. *Star Wars Jedi: Fallen Order* (Electronic Arts, 2019), Microsoft Windows, PlayStation 4, Xbox One, Stadia, PlayStation 5, and Xbox Series X/S, music by Stephen Barton and Gordy Haab.
27. Frank Lehman, "Complete Catalogue of the Themes of *Star Wars*: A Guide to John Williams's Musical Universe," https://franklehman.com/starwars/, updated February 20, 2022. Buhler identifies this as the most common Leitmotif of the entire series, one which is often identified with fate and destiny in addition to other situations.
28. See Buhler, "*Star Wars*" for an additional in-depth analysis of Leitmotif in the original *Star Wars* trilogy.
29. The scene can be viewed by the reader at https://youtu.be/5kjoOAEj17o.
30. Claudia Gorbman as cited in London, "Leitmotifs," 91.
31. OST=original soundtrack. GameFAQs, "Why is no one talking about the music? *Star Wars Jedi: Fallen Order*," *GameFAQs Message Boards*, accessed August 22, 2021, https://gamefaqs.gamespot.com/boards/240966-star-wars-jedi-fallen-order/78163722.
32. Tim Summers argues that, in games that draw from film music, players perceive these references as general media tropes rather than a one-way influence from film to game. See Summers, *Understanding Video Game Music*, 149.
33. Buhler, "*Star Wars*," 44.
34. *Star Wars: Knights of the Old Republic* (LucasArts, 2003), Xbox, Microsoft Windows, MacOS X, iOS, Android, and Nintendo Switch, music by Jeremy Soule.
35. The scene may be viewed here (note that the video creator is reading the dialogue; this narration is not from the original game audio): https://youtu.be/lO3e9KeQcug?t=60.
36. *Halo: Combat Evolved* (Microsoft Game Studios, 2001), Xbox, Microsoft Windows, MacOS X, and Xbox 360, music by Martin O'Donnell and Michael Salvatori.
37. Karen Cook, "Medievalism and Emotions," 491.
38. Karen Cook, "Beyond (the) *Halo*," 190.
39. Marty O'Donnell, interview by Chris Anderson, "Bungie's Marty O'Donnell on the *Halo 3* Soundtrack," *Wired* and Chris Anderson, April 24, 2007, https://www.youtube.com/watch?v=b1rMbTLnIbE.
40. *Halo 3* (Microsoft Game Studios, 2007), Xbox 360, Xbox One, Microsoft Windows, and Xbox Series X/S, music by Martin O'Donnell and Michael Salvatori. The scene may be viewed at https://www.youtube.com/watch?v=JvEJ6VmCR4g.
41. Caplin, *Classical Form*, 11.

42. Ethan Haimo, "Developing Variation and Schoenberg's Serial Music," *Music Analysis* 16, no. 3 (1997): 355.

43. Ibid., 351.

44. As cited in Bribitzer-Stull, *Understanding the Leitmotif*, 100.

45. *Sonic the Hedgehog* (Sega, 1991), Sega Genesis, music by Masato Nakamura.

46. As the composer, Masato Nakamura explains, "I composed melodies that kept the game tempo in mind without sounding unnatural. I also wanted to make sure that the music didn't lose its groove. After all, one of Sonic's key elements lies in speed." Masato Nakamura, interview by Takashi Iizuka, "Masato Nakamura Interview by *Sonic Central*," *Sonic Central*, May 18, 2005. https://web.archive.org/web/20081223022942/http://www2.sega.com/sonic//globalsonic/post_sonicteam.php?article=nakamura.

47. As mentioned in the introduction, to analyze segments and phrase structure I will use the grouping structure methodology of William Caplin, as presented in *Classical Form*. Although Sonic is clearly *not* a work from the classical era, Caplin's strategies for grouping motive and phrase at the smallest scale are general enough to provide us with useful tools for understanding the structure of more modern musics.

48. Huron, *Sweet Anticipation*, 195.

49. Ibid., 201.

50. Bribitzer-Stull, *Understanding the Leitmotif*, 11, 12, 95.

51. Ibid., 6.

52. Arnold Schoenberg, *Fundamentals of Musical Composition*, edited by Gerald Strang and Leonard Stein (London: Faber & Faber, Inc., 1967), 171.

Four

Retro Gaming Experiences

This chapter will explore the recent rise in popularity of "retro" games, and some of the popular musical features used within these games. Motivated partly through a change in the gaming industry to smaller, self-produced game developers, the retro aesthetic often explicitly references elements from late 1980s and early 1990s gaming. From a musical standpoint, digital instrumentation, looping structures, and pop-inspired melodies help to achieve this affect, but the pairing of these sounds with 1990s-inspired graphical styles and game mechanics also reinforces the player's sense of belonging in the past. But is this truly a replication of 1990s videogaming? Many of the analyses in this chapter will discuss how retro games invoke nostalgia through references to the digital past, but that they also pick and choose which elements to incorporate in order to achieve constructive authenticity (players' acceptance of the retro style established as conforming to their expectations of the past) without replicating some of the more problematic elements of the era.

Given that interactions between visual, sonic, and narrative components are at the heart of how players perceive game (including the effectiveness of retro gaming sound), my analyses will explore alternate modes of representation via transcription of scores, soundscape, and other elements of specific game scenes. Musical events will not be presented in isolation within the analyses, but instead will be represented in their narrative contexts since that is how players experience them. The final section of the chapter will discuss what role authenticity plays in creating an effective retro gaming experience.

THE RISE OF RETRO GAMING IN THE TWENTY-FIRST CENTURY

The start of the twenty-first century brought about significant developments in game design and gaming cultures, with the release of the Xbox in 2001 creating new competition for existing companies Nintendo and

Sega, but also including higher-speed broadband capabilities, a feature that allowed for the digital distribution of games through the release of its online Xbox Live service in 2002. While such distribution networks originally restricted access to third-party software, the launch of Valve's Steam service in 2002, and in particular its Greenlight program in 2012, opened access to third-party developers, thereby creating new means for independent game developers to distribute their content to potential audiences.[1,2]

The result was the rise of a new genre, indie gaming, a genre described by Felan Parker and Jennifer Jenson as an "increasingly visible and central part of the industry and culture of digital games." Indie gaming as a term is difficult to define precisely due to its evolving meaning over the last twenty years of gaming culture, including its function as "alternately a mode of production and/or distribution, a genre designation, a market category, a social identity, and a taste culture."[3] For the purposes of this chapter, I am interested in indie's definition as a game genre and aesthetic. As Maria Garda and Paweł Grabarczyk describe, indie games maintain a particular *feel* through a distinct set of properties: digital distribution, experimentation, small budget, low price, retro style, small file size, small development team, indie mindset, interaction with online indie communities, and the use of middleware software design tools such as Unity.[4] The result has focused on particular gaming features that position indie as an independent project ethos but which have also contributed to a growing popularity of the retro gaming aesthetic style. For example, Garda and Grabarczyk cite the revival of the 2D platformer format as the reemergence of a forgotten game mechanic, an aspect of indie gaming in opposition to large studios' focus on continually improved graphics throughout the 2000s and 2010s.[5] Many of these 2D platformers have intentionally hearkened back to other stylistic elements of their predecessors to create this retro aesthetic, including the simpler pixel-art style of older 8-, 16-, and 32-bit games but also musical elements such as looping, digital timbres, and other features to be discussed shortly. The result has been a popular aesthetic within the indie game community that strongly evokes nostalgia among players for their childhood gaming experiences, as well as a growing discussion around authenticity in respect to both graphics (namely, whether authenticity in modern gaming necessitates photo-realistic graphics—does it need to employ objective authenticity or could it be evoked in other ways?) and narrative (whether indie games' prioritization of story over cutting-edge design makes for more engaging gaming experiences, thus more effectively producing constructive and existential authenticity). I will explore these questions, but the upcoming analyses will also break them down into specific musical features to discuss how music might evoke these nostalgic, retro experiences.

RETRO AND NOSTALGIA

The retro aesthetic and its associated nostalgia have grown exponentially through the 2010s to become, as Andra Ivănescu describes in the introduction of *Nostalgia: The Way It Never Sounded,* a significant component of gaming culture in the 2020s that shows no sign of abatement.[6] Ivănescu identifies that nostalgia and the retro aesthetic, rather than being limited to retro gaming, seem to be developing into a defining feature of early twenty-first-century popular culture, a property developed by its players rather than its vendors. However, the rise of indie gaming, despite its origins as a countercultural response to large-studio game development, has also had a reverse impact on the choices of game releases and development by large studios in recent years. Studios such as Nintendo, for example, have embraced retro's popularity by re-releasing games from their back catalogues (many of which were only previously available through emulators after being discontinued on consoles), remastering earlier games such *The Legend of Zelda: Link's Awakening,* and re-releasing new versions of older game consoles such as the NES Classic Edition in 2018, an emulator featuring thirty built-in games from the original Nintendo Entertainment System (NES) that mimics the look of the original console with new hardware. These re-releases have allowed companies to reduce costs (and risk) by choosing already-developed intellectual properties proven to be successful. This is not surprising given the previously discussed trend towards sequels and media franchises in the early twenty-first century.

Other authors such as Nikita Braguinski suggest that retro's appeal and its integration with nostalgia is more complex than just a simple nostalgia effect, instead appealing to more basic aesthetic responses. Braguinski argues that some who grew up with lower-resolution formats were drawn to the return of this aesthetic, and even came to prefer these over better-resolution formats.[7] I do not believe this is the sole factor given that many players care more about the game's overall enjoyability of story and gameplay structure than adherence to a particular aesthetic, what Ivănescu describes in retro games as "a criticism of technological development viewed as evolution: their endearing pixelated worlds and characters communicating that the quality of a game and its power to create an immersive environment does not necessarily correlate with spectacular graphics and technological improvement."[8] Regardless of the reasons that players might prefer these lower-resolution formats, most would not deny that the simplicity of retro games provides a refreshing contrast to the visual complexity of some modern game design.

That said, the retro games of today are not true replications of the 1980s gaming experience, a fact that has implications on this discussion of authenticity. While retro games often use tools such as restricted graphics and colours as well as simplified sound design to mimic games of the earlier era, they are often selective in this approach; a game might use 8-bit graphics characteristic of 1980s consoles, for example, but retain a more modern soundtrack. How do we quantify authenticity in these cases? Would this picking and choosing of features break any true sense of authenticity, or is authenticity merely a matter of matching players' expectations of 1990s gaming? Ivănescu, in describing the game *Evoland 2*, suggests that these design choices are flexible:

> Much of the game's nostalgia appeal does lie in its "16-bit"-ness, in its aesthetic pastiche, but other explorations of the past it offers are much more extensive . . . in terms of gameplay, narrative and humour. These deconstruct significant elements from these games and occasionally even present unusual juxtapositions. . . . Here, music plays an interesting role, either referencing the same game/genre as the aesthetics and the gameplay or superimposing a different style.[9]

Key here are the terms *pastiche*, *referencing*, and *superimposing*. The implication is clear that there is a merge of different aesthetics, and that this does not inhibit the game's ability to produce nostalgia. However, many games in the modern retro aesthetic take a 2D adventure/platformer perspective that suggests a specific connection to early 1990s gaming cultures, including releases such as *The Messenger*, *Celeste*, and *Undertale*. Can retro authenticity refer to eras other than 8- and 16-bit gaming, or is it specifically a reference to that particular visual and aural aesthetic? Let us begin with an example that blends references to multiple eras in order to answer this question.

EVOKING RETRO IN GAMES: A CASE STUDY

Cheekily advertised as "An Insurance Adventure with Minimal Colour," *Return of the Obra Dinn* provides a good starting point for our discussion of retro aesthetics and the player's perception of authenticity. *Obra Dinn* is an indie game released in 2018 that presents a murder mystery aboard a merchant ship in the early 1800s.[10] The game was created by solo developer Lucas Pope, well known for several indie games that exploit unusual gameplay or narrative mechanics such as *Papers, Please* (2013), which puts the player in the role of a communist immigration inspector who must make life-or-death decisions about whether or not to follow ever-changing border restrictions.[11] Both games evoke a retro style, with *Papers, Please* using an early-1990s inspired colour palette, 2D pixel graphics, and gameplay interface, and *Obra Dinn* animated in two-colour graphics reminiscent of 1980s home computers (with the choice to select between screen colours modelled on the displays of early home computers such as the Apple Macintosh, Commodore 64, and IBM personal computer). Despite this reference to the graphics capabilities of systems from the early 1980s, *Obra Dinn* features a visual style that combines elements of early Macintosh retro graphics with 3D animation and also evokes another era—that of the nineteenth-century narrative setting—through the use of visual shading. As one source states,

> *Return of the Obra Dinn* is rendered in retro style—this time, beautiful black and white line drawings associated with early Macintosh graphical adventures. This style also chimes with drawings of the early nineteenth century, rendering faces that have personality, but aren't clearly identifiable. The vagueness of the imagery adds to the mystery.[12]

Pope, as the sole developer, also composed the music for both *Papers, Please* and *Obra Dinn*. Sonically, *Obra Dinn* breaks away from the 1980s aesthetic established in its visuals, with a soundtrack that includes realistic recorded dialogue and music that employs digitally sampled, realistic-sounding orchestral instruments such as cello, tubular bells, and clarinet, common in 2010s and 2020s gaming soundtracks. So does this music qualify as "retro"? Despite not adhering to the same aesthetic as the game's graphics, the answer is yes—surprising given its reliance on more modern realistic sounds. Zach Whalen and Laurie Taylor observe about this sort of merger that "this reconfiguration of the old within the new follows the logic of nostalgia that combines the past and the present in a way that can cause the past to become a fetish."[13] Thus there is an idealization at play in players' interactions with retro gaming: a desire to experience the positive aspects of past without its inconveniences.

Let us take a few moments to first outline some features of retro sound, after which we will examine one track from the game, "Soldiers of the Sea," to break this question down in more detail.

Retro Game Music Features

How might we know when sound and music features correspond to our expectations for retro sound? Scholars have begun to examine the properties of so-called 8-bit and 16-bit sound, as well as their connection to the twenty-first-century retro aesthetic, identifying several concrete features that characterize its audio. Karen Collins and Nikita Braguinski, for example, both indicate tuning that does not correspond to musical scales in equal-tempered tuning (a result of technological constraints on early gaming systems), the use of simple wave forms to create sounds with a particular focus on triangle, pulse, and sine waves, and a low number of simultaneous musical voices (typically between three and five simultaneous sounds or pitches).[14] Braguinski additionally identifies the use of temporal grids for sound programming, resulting in a limitation on the total number of pitches and durations and a lack of fine control over musical features such as dynamics and tempo.[15] Finally, both Collins and Braguinski emphasize the role of extensive musical repetition, particularly in terms of looping and formal section repetition.[16] I would like to further extend this list, as shown in figure 4.1. First, the use of simple wave forms can be expanded to an emphasis on synthesized over sampled sound (what I will sometimes term *digital* versus *acoustic* sound based on perception by their listeners). Second, musical repetition often involves very specific techniques including bass ostinato, four- or eight-bar phrase loops, and the use of repetitive large-scale forms such as verse-chorus structure, as well as a limitation on the length of musically unique segments of music.

While Collins cites a use of progressive rock and other musical forms in the 1980s and early 1990s on the part of some game developers as an intentional choice to create a unique image for their brand,[17] neither Collins nor Braguinski discuss the use of harmony or pop-evoked styles in retro games targeting the same era despite the fact that such references to pop and rock styles are quite common in retro games. *The Messenger* (2018), for instance, features a musical soundtrack by chiptune composer Eric Brown (Rainbowdragoneyes) that clearly evokes the virtuosic electric guitar styles of the 1980s with rapid sixteenth-note runs, frequent pitch bends, and natural minor mode. Created in tracker software specifically meant to imitate the 8-bit NES and 16-bit Sega Genesis styles, Brown's soundtrack clearly references earlier fighting and adventure games in the same vein, including *Ninja Gaiden* (1988, on which *The Messenger* is clearly modeled).[18] The use of natural minor mode here creates one last feature I would like to add to our retro sound features. The scales and harmonies of these games (much like the Greek-inspired games in the gameworld chapter) often use scale-degree ♭7, resulting in the more frequent use of

- Distorted tuning
- Simple wave form sounds, emphasizing synthesized ('digital') over sampled sounds ('acoustic')
- 3-5 track instrumentation
- Lack of musical nuance in dynamics, tempo, and other elements
- Extensive musical repetition, especially bass ostinato, looping phrases, and large-scale forms with recurring sections
- References to 1980s pop & rock styles
- Use of lowered scale-degree 7, natural minor mode, and flat-VII chord

Figure 4.1. Common properties of retro game music. *Source:* Figure by the author.

natural minor mode and the ♭VII chord. This avoidance of the semitone-producing leading tone helps to smooth over looping phrases, as will be discussed shortly.

Figure 4.2 gives the melody and harmonic reduction of "Soldiers of the Sea," a song that accompanies the penultimate chapter of *Return of the Obra Dinn*. Narratively, the song occurs alongside flashbacks that the player accesses via a mysterious stopwatch; in this particular chapter, the player views the boat's crew under attack by gruesome sea monsters. "Soldiers of the Sea" does not use digital timbres as might be expected given the game's focus on retro graphics. Rather, the song features a mix of orchestral strings, bass clarinet, tubular bells, and timpani to produce a decidedly realistic sound. However, despite this orchestral timbre the song does prominently include several other features of retro sound. First, the work is structured around looping, which is particularly apparent in its chord progression: after a four-bar introduction, the progression i–III–VI–v in G minor recurs, first in eight-bar phrases until bar 20 and in

Figure 4.2. Harmonic analysis, "Soldiers of the Sea," *Return of the Obra Dinn*. **Source:** Composed by Lucas Pope. Reduction and analysis by the author. Courtesy Lucas Pope.

four-bar phrases starting in bar 25, until the end of the song. Other looping also occurs, such as the melodic repetition of bars 5–12 in bars 13–20.

However, unlike video game music from the 1980s and early 1990s, the repetitions here feature substantial variation. In early 1980s games, repetitions were used to lower the amount of memory needed to store music since game designers could simply re-reference the same segment of memory when repeating a musical passage. In "Soldiers of the Sea," in contrast, the accompaniments are varied upon repetition with changes in accompaniment patterns. For example, as shown in figure 4.3, there is a change in the string accompaniment rhythm in bars 25–28, 33–36, and 46–49, with the first passage containing more syncopated rhythms, the second passage using metrically regular eighth notes, and the third passage combining the syncopated and straight eighth components. Additionally, accompaniment voicings are slightly modified to result in new

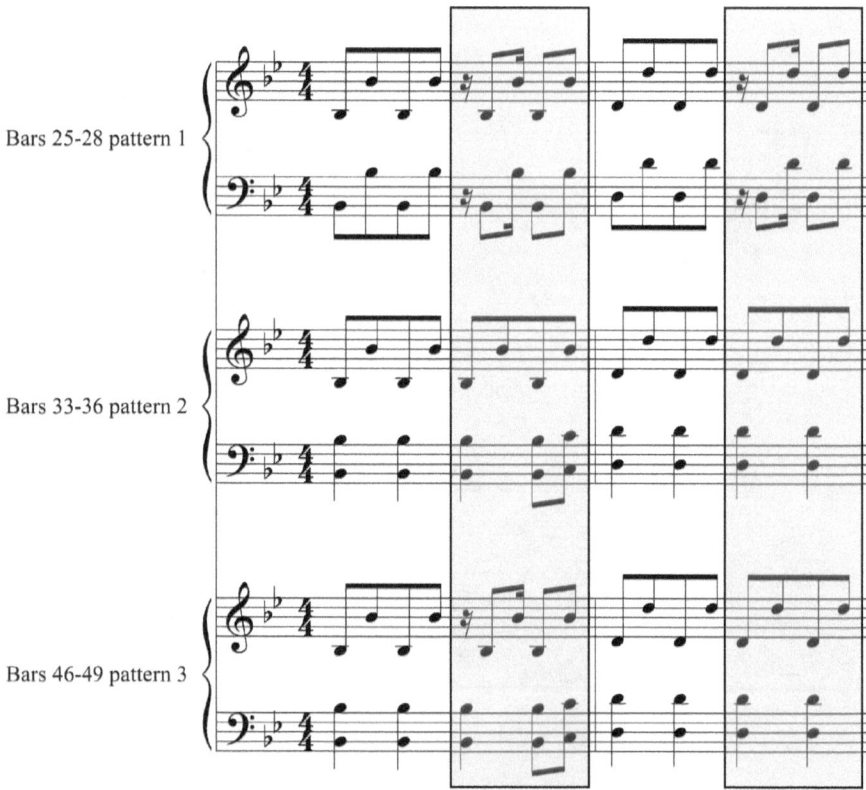

Figure 4.3. Varied accompaniment patterns, "Soldiers of the Sea," *Return of the Obra Dinn*. Differences are boxed in grey. *Source*: Composed by Lucas Pope. Reduction and analysis by the author. Courtesy Lucas Pope.

passing or anticipatory chords (indicated in brackets, such as those in bars 14 and 16 compared to bars 6 and 8), and a transposition to the new key of D minor occurs in bars 38–45. Furthermore, passages such as bar 25 and 37 act to disrupt the eight-bar phrasing through rests in the melody and a thinning of the instrumental texture, a process that creates sectionalization within the work. A similar process occurs in the final measure of the work, where rests in all instruments create a clear gap before looping back to the thinner string texture and slower harmonic rhythm of the song's beginning. Repetition and looping here are being used aesthetically to evoke a retro-like style, rather than functionally to authentically mimic the style through similar technical requirements of the system.

Extensive repetition is not the only feature that evokes a retro style in this song. The work also uses F-natural rather than F♯, thus setting the song in G-natural minor (and later D-natural minor). This means that, rather than using the tension-producing major-V and diminished-vii° chords, the song employs minor v and major VII. While raised leading-tone dominant-to-tonic progressions typically produce an effect of harmonic tension-to-resolution, the use of the lowered seventh scale-degree instead smooths out the point of harmonic looping, avoiding a clear cadential arrival. The flattened VII chord is a prominent feature of popular music from the 1970s and into the present. As Allan Moore observes, "one fundamental site of this separate identity [from common-practice harmony] is the frequent absence of a diatonic scalic leading-note."[19] Moore's observations focus on rock music, stating that the flattened seventh scale-degree in rock results in an emphasis on modal usage and alternate cadence patterns ♭VII–I and ♭VI–♭VII–I that emphasize linear bass motion,[20] but these features also occur in many video game music examples from the same era. While many phrases in "Soldiers of the Sea" conclude with a form of the dominant chord, this is a minor dominant without the raised leading tone. The result is a weakening of harmonic function as there is no implication for the music to resolve given the lack of leading tone. This, paired with a frequent emphasis on the relative major chord, B♭, obscures whether the work is major or minor and weakens the strength of the tonic arrival.

Moreover, the harmonic progression of the song strongly skews towards common-tone relationships rather than motion by fifth, yet another move away from common-practice harmonic paradigms that more typically emphasize tension and release through dominant-to-tonic relationships. As shown in figure 4.4, the repeating chord progression G-- B♭+– E♭+–D-, alternately expressed as i–III–IV–v, might best be mapped out in Riemannian space given its strong emphasis on common tones: G- to B♭+ produces a relative chord relationship (R on the figure), B♭+ to E♭+ a dominant relationship (DOM on the figure), and E♭+ to D- a two-step

L/DOM relationship (with L representing *Leittonwechsel*, a leading-tone transformation that retains the minor third of the chord while shifting the remaining note of the chord a semitone). While the additional two chords heard in the piece, A- and F+, can be mapped via similar relationships on the figure, in most cases in the song they actually participate in step-motion progressions that act as harmonic elaborations and prolongations. For example, the initial progression from G- to F+ and back again in bars 1–5 prolongs G- through step motion with its root, while the A- chord in bar 6 fills in passing motion from G- to B♭+. In either case, the song occurs firmly in a tonal, diatonic key, and the common-tone chord relationships act to avoid any harmonic effect of tension-then-resolution that might disrupt the musical looping required for the retro style.

Thus, despite lacking some features characteristic of retro sound such as distorted tuning, synthesized digital timbres, and a lack of musical nuance such as dynamics, the game and its music still manage a retro aesthetic. Indeed, as one review of *Obra Dinn* implies, one might argue that this aesthetic of looking back to the past is evoking not only the retro systems of the 1980s/90s but also the style of the early nineteenth century through a visual etching-like style that "also chimes with drawings of the early nineteenth century, rendering faces that have personality, but aren't clearly identifiable."[21] The music, with its heavy reliance on orchestral timbres, also evokes the nineteenth century through its reference to the concert hall and players' associations of pizzicato strings, in particular with classical music despite the eras not quite matching up.

Braguinski has an interesting point to make about this historical mis-mapping, a lack of authenticity in retro game sound. He states that music we now perceive as 8-bit is "often only superficially related to examples

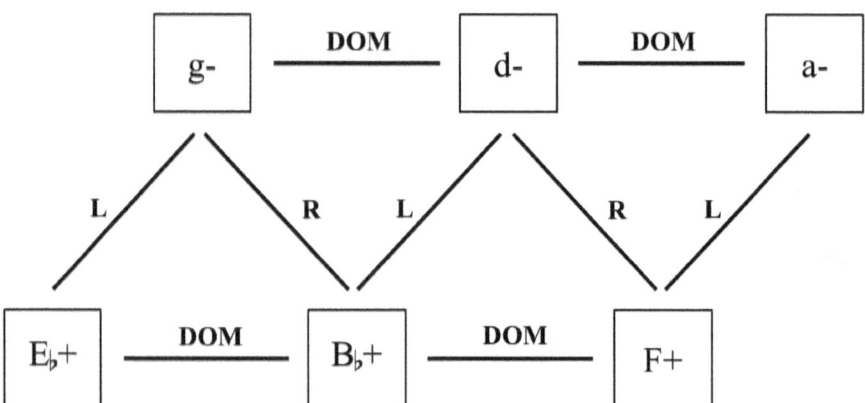

Figure 4.4. Mapping chord relationships using Riemannian space in "Soldiers of the Sea." *Source*: Figure by the author.

of previous-era audio technology and aesthetics."[22] He particularly notes that only a few of the features of 8-bit sound are required for a soundtrack to be perceived as retro, and that "everything is possible as long as there are listeners or viewers willing to accept the result as belonging to the realm of 8-bit, retro, or pertaining to classic video games."[23] Thus listeners can perceive soundtracks as authentically retro, despite other discrepancies, given one or more prominent retro-associated features. Ivănescu makes a similar observation regarding nostalgia games (a genre related to retro games), explaining that while these games reflect the historical past more broadly, there is also another level to how they are interpreted and processed by players: "History is (re)written in the process of its (re)mediation."[24] In *Obra Dinn*'s case, this reference to the past evokes not only the retro computer era but also the time period of the narrative itself, and players reinterpret the game's understanding of the past based on their own (mis)conceptions.

RETRO AND PLAYER PERCEPTION IN *UNDERTALE*

This study of *Obra Dinn* suggests that a few key features of retro style are sufficient to suggest the genre as a whole, despite other discrepancies. Other retro games with a reputation for stricter adherence to historical video game practices also feature a similar blending of modern and early styles. Let us examine one such case: *Undertale*.

Released in 2015 by independent game designer Toby Fox, the video game *Undertale* achieved massive popularity due to its unique mix of humour, nonviolent gameplay, and a retro aesthetic. Much of the game's enjoyment relies on the expectations created through the player's knowledge of earlier, similar adventure/RPG games as the game's plot centers on a child who has fallen into the monster realm and his attempts to return home by exploring the underworld. The game achieves a retro aesthetic through its use of 2D top-down gameplay, a limited colour palette and pixel graphics suggesting 16-bit style, and a simple control schema, properties that Braguinski suggests are common features of the 8- and 16-bit visual and gameplay aesthetic.[25] Several other narrative and gameplay conventions from early 1990s adventure gaming occur here: a clear duality of heroes and enemies (although as we will examine soon, one of *Undertale*'s key points is to make the player reconsider this duality), tutorial levels that present an initial exposition of plot and gameplay mechanics, and gameplay modes that switch between cut scenes, exploration, and battle. Many elements within the game satirize and subvert the expectations of adventure/RPG games and 1980s-era gaming, the time period established by this retro aesthetic, and *Undertale*'s music is no exception.

While it embodies a decidedly retro aesthetic, it is not meant to accurately re-create computerized sound from the 1980s but rather to more generally evoke a sense of retro and nostalgia.

In this section, we will examine several examples from the game that evoke retro in different ways in order to more broadly discuss player perceptions of the past: the use of 1920s flapper culture, references to 1990s popular culture, and the use of digital effects characteristic of 1980s computer sound. We will also examine how the shift into and out of retro sound impacts the player experience.

Blending Retro and Non-Retro Sounds in "Ruins"

Much of *Undertale*, like its adventure/RPG forebearers, consists of exploration mode where the player discovers new locations and characters. The first area encountered, the Ruins, is accompanied by the music given in figure 4.5, a song that begins with a more realistic piano sound—a decidedly non-retro timbre![26] However, despite this realistic instrumental timbre, the retro aesthetic of the game begins to be established here as the song's musical features create a stylistic affinity with earlier chiptune music.[27]

Repetition is a key feature, immediately established in the bass with a repeated two-bar ostinato. After a four-bar introduction of this ostinato pattern, phrases loop in eight- or sixteen-units, and a smaller melodic motif consisting of neighbour motion followed by a leap of a third also repeats in bars 12–13, 16–17, 18–19, and an extended variant in bars 14–15. The work is relatively static from a harmonic perspective, with suggestions of a [i–IV] chord progression in A♭ Dorian mode repeating every two bars, mainly articulated through the ostinato.[28] As a result, sections end not because of strong cadential phrase endings, but rather due to a petering-out of melody such as that seen in bars 19–20—where the melody thins and drops out, leaving only the repeating ostinato and accompaniment—as well as the repetition of previous beginnings (for example, the return of the melody and ostinato pairing of bar 5 in later sections, which is not shown on the figure). Thus, in a similar vein to *Obra Dinn*, the repetitive harmonic progression produces a very weak sense of harmonic shift that helps to smooth over the frequent looping processes in the song.

Bar 13 also introduces another common process in video game music design drawn from early game music practices. Upon repetition, a new instrumental layer is added to the preexisting voices in order to introduce variation while retaining the original melody and bass lines. In "Ruins," this new motive (notated in the middle staff of the figure) shifts up and down an octave as needed in order to account for registral shifts in the melody, consistently remaining the middle-register voice. It forms a new

Figure 4.5. Bars 1–20, "Ruins," *Undertale*. **Source:** Composed by Toby Fox. Reduction and analysis by the author. Courtesy Toby Fox.

repeated pattern with a new distinct rhythm (quarter–dotted quarter–eighth–eighth) but also fills in the harmony, adding G♭ as a seventh into the tonic chord in its first bar and an F in its second bar to reinforce the IV chord. Like the ostinato voice, small variations occur in the second bar of the pattern within this added voice, subtly adding variety to each repetition.

The melody of each phrase rarely ends on a scale-degree that might reinforce the tonality. D♭ is a frequent ending pitch (see, for example, bars 13 and 19, whose longer durations followed by a rest suggest a melodic cadence), but strong arrivals and longer durations on the tonic of the song, A♭, are rare to nonexistent. The result is that it is hard to hear whether the key has a tonic of D♭ or of A♭. However, the ostinato more clearly reinforces the tonality, with a recurring downbeat emphasis on A-flat. This key, A♭ Dorian mode, is similar to the natural minor mode and once again avoids the raised leading tone, thereby avoiding strong cadential arrivals (an effect that smooths over the repetition of the phrase loops, as previously discussed).

Timbrally, the opening of "Ruins" appears to focus on an acoustic piano, but this changes throughout the song, a shift that hearkens to earlier sound practices. Figure 4.6 gives a summary of the overall form of this song. At bar 21, the piano ostinato continues, but the melody is taken over by a flute in a decidedly digital timbre—not a realistic flute sampling but rather poorly synthesized reproduction of one. The shift is dramatic and clearly noticeably away from purely acoustic timbres to digitally prominent timbres, amplified by the addition of a bell-like instrument that doubles some of the melody in the highest register and drum set. These digital timbres are typical of 1980s/90s game sound, and the blending of the two contrasting timbres acts as a reminder to the player that this game is not attempting to duplicate the original 1980s/90s gaming experience, but rather is intended as an homage, an attempt to invoke nostalgia through the aesthetic of the era. Later sections continue this clear emphasis on digital timbres, with the addition of digital strings in the A2 section, digital clarinets with an extremely prominent vibrato effect in the B1 section, and a digital bass timbre in the first portion of the transition section. The return to more acoustic timbres only reoccurs in the final portion of the transition with the reentry of the piano ostinato, a preparation for the loop back to the beginning of the song.

This sense of homage is critical to the gameplay experience in *Undertale*. The game intentionally evokes moments of disruption and subversion, and musically this is the case in "Ruins." Just as the player settles in to one expectation of timbre, the sonic landscape shifts dramatically, a reflection of shifting player expectations in other components of the game. This shift into and out of the retro digital timbres mirrors changes in the graphics, for example, which change from 16-bit to a wider colour palette at later points in the game, disrupting the perception of the game as a true retro experience and instead repositioning the game as a reinterpretation of past gaming practices. The game's evocation of retro is not objectively authentic, but rather highly dependent upon constructive authenticity by meeting players' expectations of the style.

Cultural References to the Past: "Ghost Fight"

Retro in *Undertale* is depicted in ways beyond evoking the gaming technology of the 1980s. Other moments in the game make cultural references to the past, including the 1990s to early 2000s and the first half of

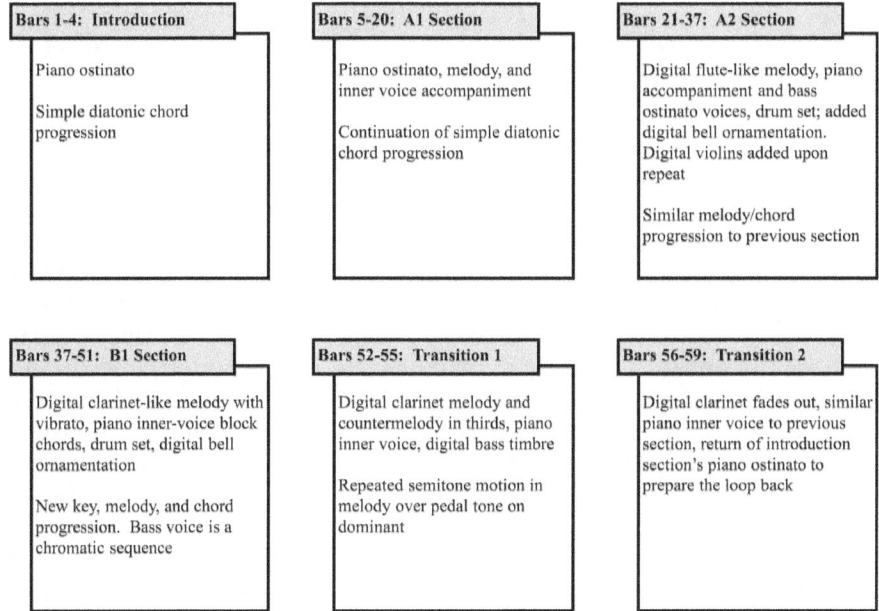

Figure 4.6. Formal sections in "Ruins." *Source*: Figure by the author.

the twentieth century, both eras that many of the game's players (as well as its designer, Toby Fox, born in 1991) have not experienced firsthand. As Andra Ivănescu discusses in *Popular Music in the Nostalgia Video Game: The Way It Never Sounded,* such references to the past are not necessarily aiming for a replication of the past, but in some cases present a reflection on the past through a modern-day lens. This perspective on the past, as in this book's earlier discussion of video games inspired by historical settings, generalizes its presentation of historical elements. "The past thus returns through the composite of an old generic universe" in a manner that makes the past more accessible to the player.[29]

For instance, many players are familiar with jazz music as a genre but not necessarily the distinctions between particular periods or styles of jazz. Divisions between Dixieland, Big Band, and Free Jazz, for instance, are outside of their knowledge base and thus players might be aware of superficial features that evoke the style, such as swung rhythms and particular instrumentations including saxophones, but not more nuanced distinctions. Jazz in video games is frequently evoked as a musical signifier of the past, particularly the American 1940s (such as in *LA Noire* or *Bioshock*, a sci-fi representation that models a failed past society), or of the *Noir* detective/mystery genre (such as in *Grim Fandango*). Ivănescu discusses several examples of these, but makes the observation that these games "draw on particular eras and their myths, but combine these images with others, creating a temporal confusion: these are not meant to be representations of the 1910s, 1920s, 1930s, 1950s, or the 1970s, but juxtapositions in which semiotic ghosts interact in unique ways."[30] Such a conflation also occurs in *Undertale* as the game does not explicitly reference a particular historical year or era, but alludes more generally to the past, including both the past of video game technology and the historical past.

Two references in the game are key examples of this point. In one scene of the game, the protagonist meets a ghost named Napstablook and is forced into battle. Napstablook's battle sequence is accompanied by the song "Ghost Fight," given in figure 4.7.[31] While this song involves digital timbres and limited voices reminiscent of late 1980s gaming consoles, its musical style, which has a decidedly jazz feel, is the more prominent feature. Swung eighth notes used in the bass and accompaniment throughout and from bar 17 onwards in the melody, and the recurring use of the C–C♯–D semitone motive (scale-degrees 4–♯4–5) to suggest the G blues scale, such as that seen in bar 2 of the melody, are prominent throughout. These two features contribute to other motives as well. For instance, the semitone motion established in the C–C♯–D motive of bar 2 recurs throughout the song but is also transposed to produce the cadential-arrival pattern F–F♯–G (accompanied by the parallel D–D♯–E motion) in bar 8. This semitone motion is also inverted and expanded, heard every two bars starting in bar 9 in the lowest voice, a walking-bass style characteristic of jazz bass performance practice. Rhythmically, the syncopated (short–long–long) pattern associated with the semitone motive in bar 8 also recurs throughout the piece, particularly at moments of cadential arrival such as in bars 24, 32, and 35–36 (not shown on the figure, but audible in the recording). Thus syncopation, motives characteristic of the blues scale, and swung eighths—three features that are perhaps some of the most obvious features of jazz to nonmusicians—act as jazz signifiers in this music. Added to this is the instrumentatio. While "Ghost Fight" still uses digital synthesis timbres characteristic of retro gaming sound,

Figure 4.7. Motivic repetition in bars 1–16, "Ghost Fight," *Undertale*. *Source*: Composed by Toby Fox. Reduction and analysis by the author. Courtesy Toby Fox.

its digital instruments are imitating clarinet and trumpet (melody), piano (accompaniment), and bass, and melodic instruments often include distortion or vibrato to approximate bent-note ornamentation used in jazz performance.

These jazz signifiers are not merely present to make an interesting piece, but rather clearly link to the visuals and dialogue presented within the battle scene, such as when Napstablook puts on a top hat and calls himself "dapper" (figure 4.8), references to popular fashion and slang from the 1920s. The combination of jazz—a style commonly associated with the first half of the twentieth century—with these visuals and narrative components evokes a generalized past, perhaps a cheeky reference to Napstablook's identity as a ghost. The *retro* in this case is thus not merely for the retro game style of the 1980s, but also refers to elements of style that blend several characteristics from an earlier time period.

A later scene with the same character makes a different retro reference. After winning the battle, the player-protagonist is able to visit Napstablook's house after they become friends, a scene shown in figure 4.9. There, the player sees various pieces of old furniture, can access several CDs, and is able to log on to a dated-looking computer, which presents the caption "(The computer's internet browser is opened to a music-sharing forum.)." At this point, players familiar with late 1990s culture might realize that Napstablook's name is meant to reference the file-sharing service Napster, founded in 1999 and popular in the early 2000s, and that

Figure 4.8. Screenshot of the Napstablook battle scene, *Undertale*. *Source*: Courtesy Toby Fox.

Figure 4.9. Screenshot of the protagonist visiting Napstablook's home, *Undertale*. *Source*: Courtesy Toby Fox.

the components of the house are consistent with this time period. The reference is anachronistic to the 1980s graphics and sound being emulated, dating slightly later. The effect is meant to produce a general retro, nostalgic feel for the early 2000s evocative of the player's (assumed) youth during that time period, rather than sticking solely to an accurate replication of late 1980s/early 1990s gaming.

Ivănescu distinguishes between *retro* games and *nostalgia* games, defining retro games as those that specifically reference earlier video games, while nostalgia games reference historical elements from any media form:

> I define nostalgia games as video games which appropriate not only the aesthetics and generic universes of media of the past, but their signifying systems, often with the potential of accessing social, political, and cultural themes through the lens (or magnifying glass) of popular culture past.[32]

Undertale does not necessarily present a critique of the past, but functions in both retro and nostalgic capacities, referencing both previous gaming technology and musical styles of the past. It is particularly self-aware of its status as a retro game (an especially strong example of what Ivănescu describes in saying "video games are particularly self-reflexive and self-aware, constantly questioning the limits of the medium and trying to push beyond them")[33] and pushes at the boundaries of players' expectations through humor but also through game mechanics. Let us examine another scene that shows this more clearly.

Analyzing Narrative Interaction though Soundscape in Undertale

One of *Undertale*'s most interesting features is the way in which retro music interacts with (or in some cases plays against) elements of both the expected game mechanics and narrative. How might we analyze this? Many video game music analyses use score transcriptions as a starting-point, as I have done with "Ruins" and "Ghost Fight." However, the most interesting features of some scenes in *Undertale* would not be included on a traditional score transcription because they cannot be represented through pitch and rhythm. Approaching this music therefore requires a rethink on the part of the analyst to interpret and represent elements such as the interaction of music with visual elements, storyline, game mechanics, and sound effects. Can new modes of representation be used to analyze juxtapositions, interactions, and subversions between these components? In this section, I will attempt an analysis of these game components through a different mode of analysis. Score transcriptions will help to examine musical features on the small scale, but I will now add new tools to help depict more interactive elements of sound, visuals, and narrative: a transcription of soundscape elements to establish inter-relationships between sound effects and music, and a timeline analysis to determine inter-relationships between sound and music, visuals, narrative, and game mechanics. No transcription or analysis can include every feature of the work, so an important aspect of these analyses will be to decide what features are most important to communicate for each scene.

Soundscape analysis, which as a methodology has precedence in both film and game music, gives us a starting point for examining these correspondences from a music-analysis perspective. Neumeyer and Buhler, for example, suggestively describes the mise-en-bande (the soundtrack) as "a kind of musical composition, and aural analysis can then be brought to bear on the sound track as a whole, its relation to the image, and its contribution to narrative."[34] Collins also sees the full soundscape as vital to our understanding of game sound,[35] and Barry Truax, in advocating for auditory scene analyses, uses language that tantalizingly suggests the sort of analysis performed by music theorists: "A major component of auditory scene analysis—sorting out complex vibrations into separate sources—is the detection of coherent patterns of those sources."[36] The transcription of musical features in conjunction with other soundscape elements will allow me to establish a timeline by which the analysis will represent the synchronization of sound with visuals, narrative, and game mechanics. In this study, I will focus particularly on timbre and moments of congruence versus incongruence, identifying when components meet or diverge from our retro gaming expectations.

Undertale begins with a storyboard cut scene that presents a brief back story: After a war between humans and monsters, which humans won and resulted in the monsters being trapped underground, a human child goes exploring in a cave and falls through to the monstrous underworld. The interactive portion of the gameplay then begins with the player-protagonist encountering their first non-player character, Flowey, for what initially appears to be a tutorial. Tutorials are a common introduction component in adventure gaming that outlines the game's controls to the player but also typically sets up the expectations of the gameworld. Oftentimes, tutorials feature a helpful non-player character who describes these controls via dialogue within the game to better integrate a purely functional aspect of the gameplay into the overall narrative.[37]

However, in *Undertale* the player's expectations are quickly subverted as the player-protagonist comes to realize that Flowey, despite the cheerful, bouncy music that accompanies him, is lying to the protagonist about the game's basic mechanics. "Friendliness pellets," which Flowey encourages the player-protagonist to collect, are actually bullets, and Flowey's recommendations are meant to harm rather than help. The player-protagonist has the choice whether to follow along with Flowey's recommendations and play the game violently, or rebel against these tactics with a pacifist approach to the game where they avoid causing harm to others—what players describe in online communities as the "True Pacifist Route." As Ivănescu states, "players may not even realize they have this choice at first (I did not) but will realise they do as they meet more monsters on their journey."[38] This choice itself runs contrary to the expectations of the genre since action-adventure games, as established in early titles such as the original *The Legend of Zelda* (1986) and *Castlevania* (1986), typically require the player to attack and kill enemies in order to achieve goals and complete quests. The subversion of this requirement sets up an expectation for the future disruption of expected adventure/RPG media tropes and forms.

Figure 4.10 gives a transcription of the music accompanying the tutorial, titled "Your Best Friend." Because the score is not able to indicate how the music interacts with other gameplay components, figure 4.11 also presents a timeline analysis of this scene that maps out its soundscape, including both music and sound effects, as well as a summary of game function components. There are clear correspondences between changes in the music and game function acting to create a sonic frame. The song is repeated, unchanged, through the initial story exposition and tutorial (0:00 to 1:02), but at the moment where Flowey reveals that he is working against the player's best interests (1:02 to 1:07)—in other words, that the game mechanics will not function as expected—the music is distorted through deceleration and transposition down a semitone.

This musical distortion and "shooting" sound effects act as cues to disrupt the progression of the expected tutorial. The entry of Toriel at 1:37, a character who assists the player, reestablishes conformity with the gamer's expectations for tutorials and is accompanied by a return to consonant music that eliminates these features of disruption. Thus, in much the same way that harmony might go through a process of stability–tension–release, the gameplay goes through a similar process. The musical tension in this gameplay corresponds to tension in timbre (dissonances) and tonal centre (the key shifts a semitone lower) and the addition of nonmusical sounds (the shooting sounds signal a threat to the player, but are more noise effects than musical sounds). There is also tension in the shift of game function itself—from the expected tutorial mode, through the subversion of this gameplay mechanic, to a return to story mode (that is, narrative exposition)—after the threat is removed.

In a later section of the game, which I will term the Dénouement scene, the player-protagonist is told the backstory of the gameworld and motivations of the main antagonist, the sum of their actions throughout the game is tallied, and they face a final reckoning with two of the major characters of the game (mini-boss Asgore and final boss Flowey). Figure

Figure 4.10. "Your Best Friend," *Undertale.* *Source*: Composed by Toby Fox. Transcription by the author. Courtesy Toby Fox.

4.12 presents an analysis of this scene that compares visuals, timbres, game phase, and narrative elements. I will begin by identifying a few significant breaks with the expectations of retro gaming at several points in this scene.[39]

First, at 6:18, the player-protagonist begins their battle with Asgore. Prior to this point, the game has used 16-bit colour but at this point colour shading is introduced (although the battle frame itself retains the pixel-graphics format), a style that visually is more suggestive of later gaming eras. The music at this moment also changes from synthesized sound to a blend of sampled and synthesized sound, creating an incongruency with the expected retro gaming aesthetic.

Next, at 13:43, everything stops, the screen turns black, and no sound occurs for nineteen seconds, an awkwardly long pause that makes the player wonder whether the game has ended. After this point, Flowey repeatedly breaks the fourth wall through metafictional references to the game state or by reloading the player to an earlier save point, creating another incongruency of retro gaming expectations. This type of disruption is not unheard of in horror genres but is unexpected in the adventure/ RPG genre, creating an incongruency in the expected narrative conventions within the genre. To give some background about how this functions in horror genres, take Sarah Bowden's description of a similar disruption effect in *Doki Doki Literature Club*:

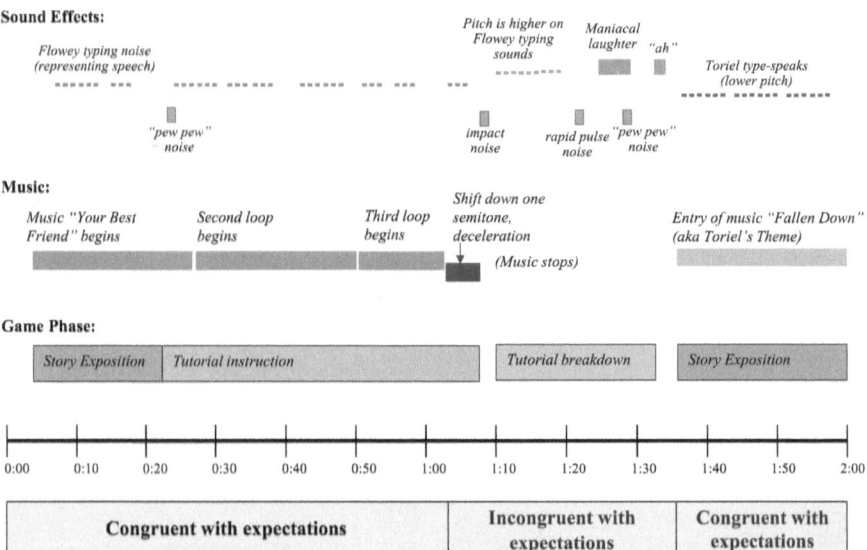

Figure 4.11.   Soundscape analysis of Flowey tutorial scene, *Undertale*. Source: Figure by the author.

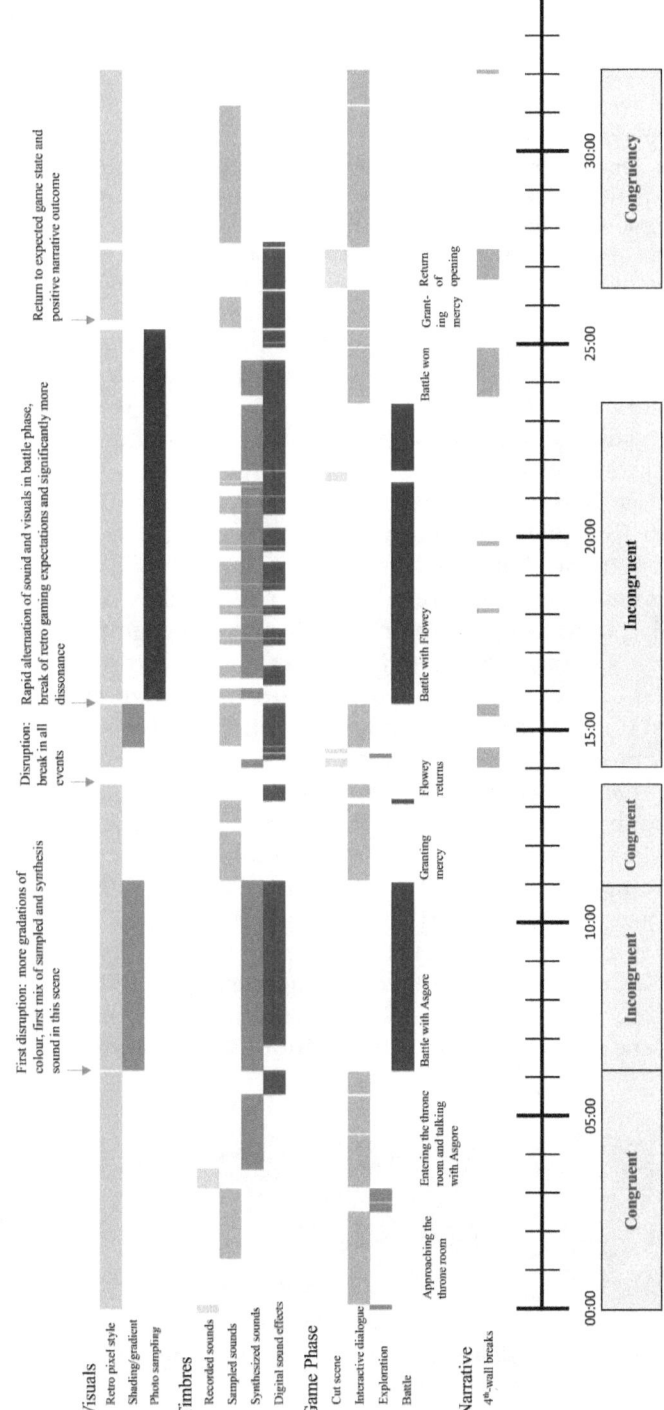

Figure 4.12. Analysis, Dénouement scene, *Undertale*. Source: Figure by author.

146   *Chapter Four*

The gamification of the operation system offers a disruption of gameplay that is necessary to continue the narrative. Forcing the player out of the game in order to continue is but one way that the player must willingly engage in disruption to continue ... [it] also function[s] as a source of horror, contributing to the player's sense of helplessness. In order to retain player interest, the sub-genre demands ingenuity that generates exponentially more terrifying gaming experiences. The game presents a low level of interactivity in regard to player interface: the player is limited to clicking to view the next panel of dialogue, a constraint that challenges the player's sense of agency. The player may only control the speed at which the narrative unfolds.[40]

The lack of control by the player is mirrored in the sound and visuals here in *Undertale*, with the too-long silence and black screen, as well as the shift to more chaotic sounds in the next battle with Flowey. The timbre profile changes to a stronger emphasis on sound effects over music, which reinforces this incongruency.

At 15:49, as the boss fight with Flowey begins, the graphics start to change from exclusively 16-bit to a mix of photo sampling (very incongruent with the retro aesthetic established thus far) and 16-bit pixel graphics, rapidly alternating between three battle scenes as shown in Figure 4.13: (a) primarily photo sampling paired with dissonant sampled music and sound effects, (b) primarily 16-bit graphics and a distorted variation of the "My Best Friend" tutorial music with distinctly digitally synthesized timbres, and (c) primarily 16-bit graphics with a blend of synthesized and sampled timbres, including prominent vibraphone on a consonant chord. The sound disrupts the aural expectations for retro/early 1990s game sound through its rapid changes, the mixture of synthesis sound with more realistic-sounding sampled music, more than five simultaneous voices, and ample use of glissandi and more complex rhythms rather than more basic temporal grids. Furthermore, the gameplay is different in each of the three components, suggesting a framing function for the sound and visuals. In battle scene (a) the player-protagonist must dodge fast-moving obstacles such as throwing stars and fire during the dissonant music; in battle scene (b) the obstacles move more slowly and predictably in

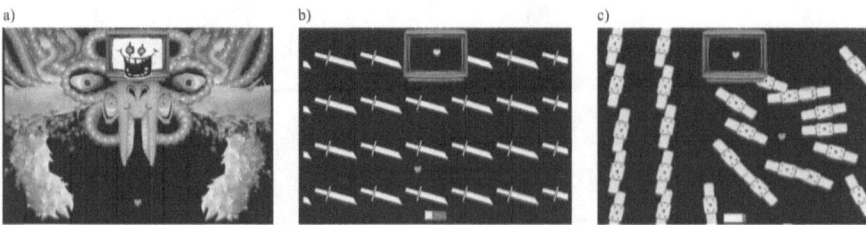

Figure 4.13. Screenshots showing three visual styles during the Flowey boss fight, *Undertale*. Source: Courtesy Toby Fox.

conjunction with the regular rhythm of "My Best Friend" as the player-protagonist tries to reach on-screen objects that call for help; and if they are successful battle scene (c)'s consonant music accompanies objects in green that increase the player-protagonist's health when reached.

Finally, at 25:32, the player-protagonist chooses whether to grant mercy to Flowey. The game "restarts," the credits play, and characters are reintroduced; 16-bit pixel graphics return; and the music returns to a focus on sampled music and sound effects, returning to the game's expected retro style.

Overall, changes between components of the scene match to significant contrasts in timbre, alternating between retro synthesized sound that meets with the player's expectations from earlier in the game and a mix of recorded sound, sampled sound, and digital sound effects that break with the overall retro aesthetic of the game. This disruption in sound matches moments of disruption in the narrative, such as the fourth-wall breaks during the Flowey boss fight, and corresponds to a higher density of game events and required player inputs. The scene is visually chaotic and sonically jarring because of rapid changes and the layering of different timbres, but the player is also overwhelmed by the significant increase in the pace of actions they must undertake compared to the earlier portions of the game. The result is an overloading of game stimuli in all dimensions that amplifies the player's emotional response to the moment of final conflict. A simplicity of graphics, sound, and gameplay occurs in areas that meet narrative expectations (congruency), while areas of more complexity in these dimensions signal a disruption to narrative conventions (incongruency on the figure). Furthermore, there is a significant break not only with retro gaming expectations, but also with the expectations of *Undertale* itself. When following the True Pacifist Route, the player-protagonist learns that granting mercy and being kind to their enemies will result in positive, peaceful outcomes. In the Dénouement scene, though, granting mercy to Asgore results in a fourth-wall break that destroys the player's saved game and results in the murder of the character the player-protagonist has just made peace with. The rapid changes between congruency and incongruency keep the player on edge, thus positioning retro authenticity not as a default state that the game is aiming for, but rather as a metric for how the player should engage with the narrative components of the game.

FINAL THOUGHTS ON THESE ANALYSES

So is *Undertale* an authentic retro experience? In many ways the question is irrelevant. The game is not meant to be an accurate duplication of a historical original, but rather is intended to be both a critique of established modes of gameplay as well as an homage to the 16-bit games of our (presumed) youth. As Ivănescu explains in her exploration of nostalgia, retro games such as *Undertale* often take a perspective of reflective nostalgia that filters the past through modern-day social and cultural lenses, and thus many retro games "are used as opportunities to critique aspects of games of the past, providing commentaries and opportunities for reflection on the part of their players."[41] *Undertale* is certainly a prime case of this, with its pacifist-centered approach acting as a particularly strong subversion of gaming conventions, undermining its own status as a retro game by repeatedly reminding the player about the discrepancies between the genre's expectations and the reality of the game. The effect of these disruptions is to shift the game away from nostalgia towards a gameplay experience that questions everything we have learned about gameplay conventions over the last thirty years, with constructive authenticity providing the necessary framework of expectation.

Finally, the effect of these disruptions is to shift the game away from nostalgia towards a gameplay experience that suggests the uncanny, but which also echoes the narrative's themes of pacifism, forgiveness, and self-agency. Heather Osborne, in relation to a different series of games, observes:

> The tension that arises between the familiar and unfamiliar . . . shifts nostalgia towards uncanny horror. As the games' narrative content diverges from the initial nostalgic impressions created by their "8-bitness," the games' horror genre begins to show. Gradually, unsettling details develop that trouble the games' superficially simple narratives. Gameplay becomes disquieting as the player's agency is disrupted or foreclosed. Evoking the uncanny breaks through the idealization of the past that nostalgia requires. The comforting, familiar past becomes more difficult and more distant. In this way, players' nostalgia for the familiar past is both conjured and foreclosed by the games. Rouse notes that in video games, "horror can be used to introduce unique gameplay mechanics based on this altered reality" (2009: 17). The sense of uncanny dread is reinforced by the games' mechanics, as the gameplay enhances, echoes, or mirrors the games' narrative themes. In this way, players are denied the "basking" in the past that Suominen (2008) associates with video game nostalgia. Instead, as the games progress, players must confront the horror that underlies the narratives and game mechanics.[42]

While *Undertale* is not a horror game, the disruption and lack of control (particularly in the Dénouement scene) share some very similar elements: a shift from a simpler narrative to a more complex one that brings in player agency as a major theme, for instance, and a mental shift away from basking in the past towards understanding the gameplay as a more critical commentary of players' ludic assumptions as the story progresses. The more basic elements of retro sound are certainly important in establishing the gameworld—the features I have highlighted here steer the player towards a particular frame of reference in terms of genre, gameplay parameters, and narrative expectations at the beginning of the game—but more and more as the game progresses it becomes clear that these games are not trying to replicate games of the early 1990s, but are instead using them as a springboard for larger questions about player interaction. As Collins explains, "players have some control over authorship (playback of audio) that is of particular relevance."[43]

While *Obra Dinn* does not critique gameplay conventions in the same manner as *Undertale*, it does present a challenge of how effectively a story can be told with minimal programming resources. In this case, the music acts as a tool to provide texture to the extremely simple visual universe through instruments that suggest the nineteenth century, more realistic timbres that go beyond the 1980s-inspired graphics, and harmonic and looping structures that are both reminiscent of 1980s gaming but also more complex and rich than computer music of that era. It also acts as a framing device as the game switches between modes of gameplay, with music accompanying scenes where the player accesses moments of violence from the past but that is notably absent from the free exploration mode of the present. In both games, players perceive the musical features of retro as an element that establishes the genre of the game, but with secondary functions that become reinforced (much in the same way as Leitmotif) through contextual repetition as the game progresses. The authenticity is thus to the game itself as well as player expectation and emotional reaction, suggesting constructive and existential authenticity over an objective authenticity based on historical accuracy.

NOTES

1. Pierson Browne and Brian R. Schram, "Intermediating the Everyday: Indie Game Development and the Labour of Co-Working Spaces," in *Game Production Studies*, eds. Olli Sotamaa and Jan Švelch (Amsterdam: Amsterdam University Press, 2021), 87.

2. Maria B. Garda and Paweł Grabarczyk, "Is Every Indie Game Independent? Towards the Concept of Independent Game," *Game Studies* 16, no. 1 (October 2016), http://gamestudies.org/1601/articles/gardagrabarczyk.

3. Felan Parker and Jennifer Jenson, "Canadian Indie Games between the Global and the Local," *Canadian Journal of Communication* 42, no. 5 (2017): 868.

4. Garda and Grabarczyk, "Is Every Indie Game Independent?"

5. Ibid.

6. Andra Ivănescu, *Popular Music in the Nostalgia Video Game: The Way It Never Sounded* (Cham, Switzerland: Palgrave Macmillan, 2019), 14.

7. Nikita Braguinski, "The Resolution of Sound: Understanding Retro Game Audio Beyond the '8-Bit' Horizon," *NECSUS: European Journal of Media Studies* 7, no. 1 (2018): 114.

8. Ivănescu, *Popular Music*, 14.

9. Ibid., 36–37.

10. *Return of the Obra Dinn* (Lucas Pope and 3909 LLC, 2018), MacOS, Microsoft Windows, PlayStation 4, Nintendo Switch, and Xbox One, music by Lucas Pope.

11. *Papers, Please* (Lucas Pope and 3909 LLC, 2013), Microsoft Windows, MacOS X, Linux, iOS, and PlayStation Vita, music by Lucas Pope.

12. Colin Campbell, "*Return of the Obra Dinn* Is a Superb Murder Mystery Game," *Polygon* (October 19, 2018), accessed May 25, 2020, https://www.polygon.com/reviews/2018/10/19/18001242/return-of-the-obra-dinn-review-windows-pc-mac-steam.

13. Whalen and Taylor, *Playing the Past*, 3.

14. Braguinski, "The Resolution of Sound," 108; and Karen Collins, *Game Sound: An Introduction to the History, Theory, and Practice of Video Game Music and Sound Design* (Cambridge, MA: MIT Press, 2008), 21–25.

15. Braguinski, "The Resolution of Sound," 108.

16. Braguinski, "The Resolution of Sound," 109; and Collins, *Game Sound*, 19, 26–28, and 31.

17. Collins, *Game Sound*, 43.

18. The VGMbassy Podcast, "Episode 13—The Messenger with Composer Rainbowdragoneyes," accessed September 5, 2021, https://thevgmbassy.com/2018/10/12/episode-13-the-messenger-with-composer-rainbowdragoneyes/.

19. Allan Moore, "The So-Called 'Flattened Seventh' in Rock," *Popular Music* 14, no. 2 (1995): 187.

20. Ibid., 193.

21. Campbell, "*Obra Dinn*."

22. Braguinski, "The Resolution of Sound," 107.

23. Ibid., 111.

24. Ivănescu, *Popular Music*, 55.

25. Braguinski, "The Resolution of Sound," 111–113.

26. Many thanks to Joey Cantarutti, my student research assistant, for his assistance with the transcription for "Ruins."

27. Gameplay of this scene is available to watch at https://youtu.be/ba0tnpWBZw8?t=242.

28. Note that other voices sometimes introduce notes that work against the ostinato harmony; for example, the G♭ in the inner voice in bar 17. This, however, does not substantially alter the perception of this harmony as the repetition has already established the harmonic progression's link to the bass voice. Additionally, sevenths have not been included in the harmonic analysis as these vary upon the repetition of the ostinato progression.

29. Vera Dika, *Recycled Culture in Contemporary Art and Film* (Cambridge: Cambridge University Press, 2003), 10, as cited in Ivănescu, *Popular Music*, 16–17.

30. Ivănescu, *Popular Music*, 109–110.

31. The official soundtrack is available to hear at https://tobyfox.bandcamp.com/album/undertale-soundtrack.

32. Ivănescu, *Nostalgia Video Game*, 17–18.

33. Ibid., 31.

34. Neumeyer and Buhler, *Meaning and Interpretation of Music in Cinema*, ix–x.

35. Collins, *Playing with Sound*, 4.

36. Barry Truax, "Acoustic Space, Community and Virtual Soundscapes," in *The Routledge Companion to Sounding Art*, eds. Marcel Cobussen, Vincent Meelberg, and Barry Truax (New York: Routledge, 2017), 257.

37. This scene is available to watch at https://www.youtube.com/watch?v=O6RNVHQOjJ8&t=37s.

38. Ivănescu, *Popular Music*, 32.

39. The scene and its accompanying analysis can be viewed at https://youtu.be/uSA50_hRHZY.

40. Bowden, "Not Suitable for the Easily Disturbed," 12.

41. Ivănescu, *Popular Music*, 14.

42. Heather Osborne, "8-Bit Nostalgia and the Uncanny: Horror as Critique in Twine Games," *Horror Studies* 9, no. 2 (October 1, 2018): 220.

43. Collins, *Game Sound*, 4.

Conclusion

I would now like to discuss the implications for player perceptions of authenticity. However, a key point I would like to make beforehand is that music alone does not create feelings of authenticity, but rather player understandings of authenticity are formed by their reception of sound, visuals, narrative, play, and the interrelations between these phenomena. While this has been modelled in our four game functions of *cueing, texturing, evoking,* and *framing,* as well as in many of the analyses in this book (such as the analysis of *Undertale* in our retro gaming chapter, with its focus on how sound and graphics shift in relation to gameplay mode), I would argue it deserves an even stronger emphasis.[1] The blend of these four elements and their resulting effects on crossing the magic circle boundary is what creates believability for the player in the game experience.

WHAT IS AUTHENTICITY?

And so the question returns: What is authenticity? In chapter 1, I defined authenticity as largely determined through player expectations, an observation that was supported by analysis throughout the subsequent four chapters. I also established some vocabulary for defining authenticity, using the trifold division of *objective authenticity,* which responds to the witnessing of historically accurate replications; *constructive authenticity,* which responds to the fit of game events with player expectations and intertextualities; and *existential authenticity,* which responds to player responses, particularly emotional responses. But what exactly have the analyses shown in terms of how these authenticities co-exist? Is it merely a fluke of vocabulary that we use the same term for all three phenomena, or is there a link between how players understand their interactions?

I propose the following points, modelled in figure 5.1.

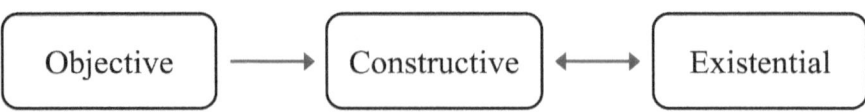

Figure 5.1. One possible model of authenticity. *Source:* Figure by author.

Objective Authenticity Can Inform Constructive Authenticity

In the examples we have seen, accuracy to history has been highly dependent on the player's understanding of history. In chapter 1, for example, my examination of games based on Ancient Greek themes demonstrated that the musical criteria that most players associated with this place and time were, for the most part, anachronistic substitutes for actual historical musical practices. On the other hand, Ubisoft made a point of integrating newer archaeological research in *Assassin's Creed: Odyssey* that impacted the design of their visuals within the game (for instance, colouring statues and temples rather than leaving them the stone white typically associated with Greek ruins). The objective, research-proven facts were selectively chosen for inclusion in the game likely based on which factors the game designers thought would best represent Ancient Greece to (and best be known by) their players, thereby texturing the gameworld.

Alexandra Wilson argues that audiences are aware of such anachronisms in film, but I am not convinced that the same is true in games.[2] Fan comments on YouTube and Reddit, such as those on the authenticity of Greek music that I previously cited in the gameworld chapter, suggest that players consider the music to be accurate reproductions of period music. While many of the texts of the sea shanties were derived from classical texts from Ancient Greece, as newly composed songs the music cannot claim the same direct connection to history since they are mediations, filtered through the imagination (and admittedly research) of their composer. Wilson's study acknowledges this mediation effect, even citing it as a recent shift in film: "Filmmakers' motivations for using musical anachronisms vary . . . the recent shift of emphasis is striking, moving from an aspiration for musical authenticity . . . to an aspiration for 'relatability,' a neologism originally used in daytime television that is increasingly being used as a yardstick by which to measure cultural worth."[3] She speculates that this intentional use of anachronism has occurred in order to better allow viewers to self-identify with the characters and situations presented, an observation that suggests an existential and constructive approach to authenticity based more on emotion rather than any strict adherence to historical accuracy. She also suggests that certain music is used for its cultural capital and prestige rather than for its original historical connotations, once again implying a form of mediation through our modern-day lens.[4] Andra Ivănescu identifies a similar mediation in her discussion of musical appropriation and argues that players need not know the exact source of the original music used in order for musical borrowings to be effective.[5] In her analysis of *LA Noire*'s music, for example, she further discusses how the game's music establishes its post–World War II noir universe and explains that our understanding of its music is not historically accurate, but rather acts as a musical shorthand: "The point is to create a sense of authentic noir through an almost mythical 'film noir sound' which has more to do with semiotic shorthand than with what classic noir sounded like. And 'noir sound' is, as David Butler notes, jazz."[6] The anachronisms and conflation of musical styles are intentional as players are better able to recognize the shorthands than the original source materials. William Gibbons makes similar observations, noting that "the kinds of shorthand codes that composers use to indicate nationality or ethnicity need not necessarily conform to *actual* musical practices—they just have to be close enough to meet audience expectations. . . . Historical accuracy is less important, in other words, than conforming to players' expectations of what old music sounds like."[7]

While Tim Summers discusses the reverse relationship—that knowledge acquired through the game in turn informs gamers' understanding of history *outside* of the game—he also appropriately discusses that this reverse-direction information flow is problematic as it creates false versions of history as well as cultural stereotypes (such as in the discussion of *Samba de Amigo* in the trope chapter).[8] Gamers, in many cases, are unaware of their imperfect knowledge of history and the fact that they create new cultural constructs from these distortions. In terms of true objective authenticity, the flow of information is thus unidirectional. Objective authenticity may contribute to the player's sense of constructive or existential authenticity, particularly if they are historically informed, but the reverse is not true (or, at least, it no longer remains as objective authenticity if impacted by constructive elements, but instead becomes its own version of constructive authenticity). Objective authenticity may be thus understood as equivalent to accuracy.

Objective Elements Act as Signifiers for the Constructive Universe

How do objective and constructive authenticities interact? The one-way flow from objective to constructive mode often involves historically inspired music, architecture, costume, and plotlines as *signifiers*, meant to create clear associations with time and place that help to establish narrative and gameworld through texturing. But despite being modeled on fixed historical references, such mappings are not consistent because of variances in players' knowledge and memories as well as their abilities to recognize and discriminate elements of sound and music. A gamer who is also a performer of British folk music, for instance, might know the song "John Barleycorn" from a local pub performance before encountering it in *Stronghold 3*, and thus might attune to modern-day performance techniques and interpretations of the lyrics when hearing the song rather than perceiving the game's intended positioning of the song as a historical signifier of eleventh-century Britain. Aficionados of Scottish culture might recognize the lyrics from Robert Burns's 1782 setting of the poem. English literature scholars might instead know Jack London's autobiographical work of the same name from 1913 and understand it as a reference to London's struggles with alcoholism. A historian might know all of these references and be able to contextualize the song's long-standing connections to themes of alcohol and misfortune. The references are inconsistent because human experience is inconsistent. Iain Hart explains this form of communication as a negotiation between player and game, depending as much on the player as on the constructed music and gameworld.[9] As he describes, music thus functions as a signifier in both concrete and abstract ways:

> We can address meanings on the terms in which they are transmitted—describing them with words as humans are wont to do, but not assuming they will map neatly or even consistently to a set of words or concepts. This is quite handy when we look at music, which is a largely non-verbal and abstract medium. . . . It becomes especially useful when we look at music in video games, where music can be directly associated with events, characters and areas, even while maintaining some level of abstraction from these things.[10]

Hart directly references one of the means of creating authenticity referenced earlier in this book: thematic association with events, characters, and areas in video games. As I examined in that chapter, signifiers might be derived from materials outside the game in question (such as how players perceive musical themes between sequels), but in other cases signifiers are established and reinforced within the game itself via strategies of motivic development (such as Leitmotif in chapter 3's *Halo 3* example).

Constructive authenticity is thus established when a game meets players' expectations based on their previous knowledge. This might reference highly specific elements (such as in the *Super Smash Bros.* series, where characters that cross over from preexisting franchises such as *Final Fantasy* and *Super Mario* are accompanied by the musical themes from those earlier games), but in other cases might reference general knowledge of a genre. As I examined in the tropes chapter, Fantasy games have an expectation of magic, a pseudo-medieval setting including specific musical elements such as plainchant and modes, and a narrative structure that focuses on the hero's quest. But other genres equally set up such expectations. Survival horror games, such as the *Metro: Exodus* example in the gameworld chapter, typically employ a soundscape with an overall lower volume, frequent dissonances, and audio effects that overemphasize properties of the virtual space (such as reverb to indicate large, abandoned spaces) in conjunction with visual aspects such as low lighting, a sparsely populated gameworld, and frequent combat. These associations are made through their recurrence in games, films, and television, building a collective cultural understanding of genre.

Constructive Authenticity Is a Prerequisite for Existential Authenticity

Constructive authenticity and existential authenticity, on the other hand, seem to be closely related. In chapter 2's study of tropes, I noted that player expectations were formed from previous cultural knowledge; as Matthew Bribitzer-Stull describes, much of our understanding of musical meaning is derived from

> musical association as a currency of communication, a portion of the vast number of shared memes that unite and—in part—define a culture or community. In fact, we invoke them every day, often without conscious realization of the communal storehouse we access and expect others to access as well.[11]

He suggests that our understanding of meaning is created through both cultural competencies and individual experiences, with one filtered through the other. While players certainly appreciate the beauty and craft of many games' musical scoring, the attachment of meaning and emotion to those musical works is often closely tied to narrative.

Our emotional buy-in to the game is predicated on the game meeting our expectations for authenticity in some sort of way. In chapter 2's examination of *Shadow of the Colossus*, for example, the player expected mournful, sad music upon the imagined death of their equine companion, and the soundtrack mirrors these expectations with slow, minor-mode, and untexted vocal counterpoint rich in suspensions—features that the player recognizes from other media texts with similar emotional backdrops. In that example historical elements coloured the musical interpretation through references to baroque counterpoint and requiem style, but similar emotional connotations can occur without the historical referents. Take, for example, the carousel scene from *Detroit: Become Human* (2018).[12] The game presents a near-future where sentient androids struggle for freedom and acceptance from the humans around them. In this scene, the android couple Kara and Luther panic as strangers attempt to enter the building where they are sheltering with the child they are protecting, Alice, but then they discover these strangers are actually former amusement park androids hiding from humans. The amusement park androids warmly welcome the family trio and then activate a nearby carousel to provide Alice with a moment of respite from the family's long, difficult journey. The scene shows Kara struggling with feelings of relief and happiness at Alice's chance to finally experience joy, but Kara's facial expressions also suggest that she is experiencing introspection and sadness that the family has had to experience so much conflict and violence. Many of the musical cues are similar to the "Agro Falls" scene from *Shadow of the Colossus*. As shown in the reduction given in figure 5.2 and heard in the video clip, "Carousel" is also in a minor key and a slow tempo, features that Katherine Isbister identifies as frequently producing sadness and anxiety in the player.[13] The song extensively uses added dissonances, with added notes in bars 2, 4, and 6 that create dissonant intervals resolving to consonances every second bar. A similar process happens in the second phrase (bars 9–17), but this time replaced with suspensions, making the similarity to "Agro Falls" even more apparent. These elements function as topics that combine to create the trope/media trope of loss, regret, and introspection, and players who have played *Colossus* may remember this mapped association during their gameplay of *Detroit: Become Human*, creating an intertextual association.

Instrumentation is also a factor in the perception of this trope/media trope within *Detroit: Become Human*. The song features string orchestra for the first eight bars and then adds piano and glockenspiel to the orchestration in the second phrase, instruments often topically associated with emotional richness (such as in *Halo* 3's Cortana theme) and magic or wonder, respectively. Further repetitions of the theme add a continuous eighth-note pulse in the string accompaniment as the tone of the scene shifts from threat to joy, changing the original slower tempo to one with more momentum to mirror the emotional tone. Players have cited the interaction of music and narrative in *Detroit: Become Human* as particularly effective in creating emotional engagement, with one user describing the musical setting as "actually really fitting and elevates the emotional impact of his scenes."[14]

Figure 5.2. Bars 1–17, "Carousel," *Detroit: Become Human.* **Source: Composed by Philip Sheppard. Reduction and analysis by the author. Original music copyright Sony Interactive Entertainment and Quantic Dream.**

We experience existential authenticity's emotional impact in these games because of the interactions between our expectations of music, visuals, and narrative. While players might subconsciously experience the emotional impact of game scores without explicitly thinking about any external references, the collective knowledge of their cultural and personal experience frames their ability to emotionally engage with music. In order for us to have the "proper" emotional response, the music must match our expectations. If, instead, we had heard a fast-tempo, major-mode song with high-register instruments, we would be left wondering about the apparent disconnect between narrative events and musical parameters. Is the music mocking us in some particular way? Is it satire? Is it meant to be a philosophical commentary on the nature of death and loss? David Huron states that this incongruence between perception and expectation can generate a different sort of emotional reaction:

> In some works, the thwarting of expectation may be sustained throughout the music. If the music is constructed in an appropriate fashion, listeners can experience a nearly total predictive failure.... In these circumstances the listener will be unable to reconcile the actual events with any existing perceptual schema. The psychological consequence is that the listener will experience a relatively high degree of stress and discomfort.[15]

This can manifest as stress and discomfort, as Huron observes, but it can also provide a rich source of humour as well. As an example, take the instrumental introduction to "Sloprano," an aria sung by villain The Great Mighty Poo in *Conker's Bad Fur Day* (figure 5.3).[16] The game, an early 2000s satirical take on adventure games, features several pop culture and gaming references as well as several recurring crude jokes, but in this case the music presents a blatant parody of the Toreador Song from *Carmen* ("Votre toast, je peux vous le rendre"; figure 5.4).[17] Summers suggests that "in order for such satirical humour to be successful, the domain of reference to an original must be quickly and precisely established.... The knowingly derivate nature of the theme articulates the humorous satire in play" and that is certainly the case here.[18] The first bar begins with a similar opening grace-note run, the accompaniment features a similar bassline, and the triplet motive in bar 8 is practically an exact quotation (albeit in a different key) to that of *Carmen*. Additionally, the orchestral accompaniment and baritone register present a similar instrumentation to the original aria. However, the gameplay setting is dramatically different than we might expect from *Carmen*. Rather than the martial brass topics and strong accented melodic line combining to create the trope/media trope of the heroic bullfighter, in *Conker* these musical elements are perceived as humorous because of their incongruity with the scene at hand, a confrontation with a literal shit monster dripping in disgusting layers of ooze (figure 5.5). A match of expectation might establish

Figure 5.3. Bars 1–8, "Sloprano," *Conker's Bad Fur Day*. *Source*: Composed by Robin Beanland. Reduction and analysis by the author. Original music copyright Rare.

Figure 5.4. Bars 1–8, "Votre toast, je peux vous le rendre" (Toreador Song), *Carmen*. *Source*: Composed by Georges Bizet. Public domain. Analysis by the author.

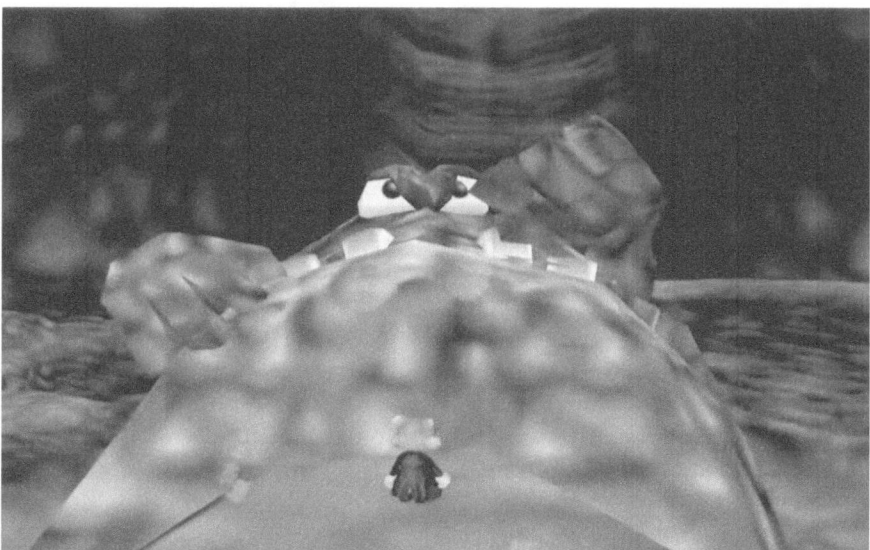

Figure 5.5. Screenshot, encounter with The Great Mighty Poo, *Conker's Bad Fur Day*. **Source**: Reproduced by permission of Rare.

the player's sense of constructive authenticity to generate basic emotions such as sorrow, joy, or pride, but the disconnect between the expectations established by the musical reference and the game's crude poop references creates surprise and humour instead.

Thus existential authenticity—that is, the player's perception that their feelings in response to the game fit within the context established by the game—is dependent on their expectations established through constructive authenticity. The scene would still be humorous because of its scatological humour without the player's previous knowledge of the Toreador Song, but the *musical reference* would not work quite as effectively. While players may not know the exact source of the aria, they will likely know from other contexts that it is an operatic aria, attaching the social codings of opera as serious and high-brow (thus amplifying the joke).[19]

Existential Authenticity Can, Conversely, Colour Our Understanding of Constructive Authenticity

Finally, the player's emotional understanding of the game can greatly impact the actions they take in game as well as their understanding of intertextual referentiality. Games play on their emotional state, often with a strong narrative component that links to further audio-visual elements. As a case study, let us examine music as a tool for communicating the impact of war. When we think of violence in video games, we most commonly imagine first-person shooters, whose casual approach to violence includes driving music meant to propel the player forward. First-person shooters often feature epic-style, full-orchestra soundtracks, as discussed in a previous chapter. However, a growing genre of games instead focuses on a different approach to war and conflict, instead centering the effect on innocent civilians rather than emphasizing violence as a means to an end. Unlike first-person shooters, the point of these games is not to eliminate opponents or to be successful in battle, but rather to engage with moments of decision-making that dramatically influence the outcome of the game. With significant elements of choice, morality, and consequence built into the game narrative, the player engages intellectually and emotionally with these games very differently than with other genres. The music of these games strongly contributes to the overall player experience through many of the components we have now discussed. Texturing in particular is a strong influence, creating a believable gameworld that the player emotionally connects to, but the evocation of emotions (especially regret, grief, and sorrow) through our three authenticities also plays a strong role in supporting the narrative. These effects are achieved through cultural reference via objective and constructive authenticities, musical structure, and the thwarting of musical expectations, but also some less tangible features.

Several scholars suggest that games can provide insight into our moral selves, despite their clearly fictional settings. William Cheng, in *Sound Play: Video Games and the Musical Imagination*, argues that the conversation around violence in video games suggests "basic questions about whether games and gameplay are really as virtual as they appear—about how our engagements with gameworlds (and the sounds in there) speak to who we are and our values out here."[20] Cheng overall explores whether players experience a true separation of fantasy and reality when gaming, or whether an "in-between," as he terms it, is a more accurate reflection of the gaming experience. He argues that music in games can act as a mirror reflecting the player's own internal thoughts.

Scholars are well aware of the potential for games to speak on social issues. Mary Flanagan, for instance, in *Critical Play: Radical Game Design*, identifies that:

> There is a fast-growing collection of computer-based games designed to educate on matters relating to environmental concerns, human rights abuse, worker's issues, land use, and other social ills. These games are often created to address real-world issues or to raise awareness and foster critical thinking. Both categories of games integrate real-world data and stories, focus on education and public opinion, and aim to provide an alternative to existing media on such issues.[21]

However, Flanagan qualifies this by stating that these games are "artist-produced computer games" that "take an 'outsider' stance in relation to a popular, commercial games culture."[22] I disagree with this. Several examples of popular, commercial games speak directly to the social issues that Flanagan identifies (including *Detroit: Become Human*, as previously mentioned). As games that discuss social issues, these inherently play on our empathy and emotion, and music plays a large part in how they communicate.

More productively for the sake of this discussion, Flanagan cites Marshall McLuhan on the power of such art forms. While McLuhan's writings obviously predate the emergence of music in video gaming, it nonetheless helps to identify gaming's power of communication. He states that "art, like games or popular arts, and like media of communication, has the power to impose its own assumptions by setting the human community into new relationships and postures."[23] Given that McLuhan mentions both media and game, it is not a stretch to extrapolate that this statement could equally apply to video games. These new relationships and postures allow the player to experience connections between narratives and music. In looking at video game music and its structure, Katherine Isbister argues that video games have the unique properties of choice and flow that generate a sense of responsibility, consequence, and more importantly empathy in the player.[24] Musical features can thus suggest empathy and link to player perceptions of authenticity, the emotional landscape of the game narrative, and immersion.

Papers, Please, an indie game released in 2013 by developer Lucas Pope, is an apt game to consider for this conversation (and one that I alluded to during my earlier analysis of *Return of the Obra Dinn*). Pope, who as previously mentioned also composed the music for the game, combined puzzle and simulation elements to explore the experiences of a border crossing agent in an implied Eastern-bloc Cold War country. Players must verify travel documents, interrogate travelers, and keep track of ever-changing regulations regarding entry requirements. From a moral standpoint, the gameplay allows the player choice on a variety of issues. They may choose whether to support their rigid government, or whether to covertly assist a dissident group, and they may also decide on whether or not to follow the rules based on their sympathy for other characters or even for personal benefit. A major influence on these decisions is to preserve the well-being of the player's in-game family by having enough daily income to feed, house, and keep them warm. The player's resources become more meager as the game progresses and unexpected events impede their ability to earn income. These restrictions create an ever-tightening noose that quickly involves the player emotionally. The limited number of actions possible in the game and the sheer monotony of the protagonist's daily life shifts the focus to conversations with other characters.

The main theme, "Glory to Arstotzka," plays at the end of each workday when the player-protagonist receives the results of their daily work and makes decisions about the impact on their family.[25] This song in many ways represents the repetitiveness of the player's daily work. First, this repetitiveness is mirrored structurally. The three main melodic sections are each repeated, and the end of the work returns the music to the start of the piece, creating a (potentially) never-ending loop. The theme features significant repetition in its bass line (shown in Figure 5.6a), alternating between an implied tonic and dominant within each tonal area, a move that can be understood as a larger-scale prolongation of each harmony through the alternation of its root and fifth. This harmony features a very simple chord progression, primarily focusing on a tonic-to-dominant motion prolonging D minor as well as a similar alternation and prolongation in the relative key of F major for the first half of the song. In the second half (a portion of which is given in part b of the figure), the song repeats the progression i–v–VI–iv–i three times to conclude with a typical tonic–subdominant–dominant–tonic cadential pattern. There is not much interesting here, and that is somewhat the point; the repetition of the same patterns, over and over, creates a predictability for the player.

Figure 5.6. Repetition in "Glory to Arstotzka," *Papers, Please*. *Source*: Composed by Lucas Pope. Reduction and analysis by the author. Courtesy Lucas Pope.

Nothing new is being evoked, and nothing the player does changes the musical playback. There is a futility here that the player perceives at an emotional level before understanding any referential rationale for its presence (although given the implied Cold War–era communist setting they might pull from their historical knowledge to assign it meaning in that context).

However, other musical settings in the game are more active and less repetitive, suggesting a different emotional impact. The "Death Theme," for instance, has features that both unsettle the listener and suggest the unfortunate consequences of the player's choices throughout the game.[26] As seen on figure 5.7, the bass line of the "Death Theme" reverses the direction of the main theme's bass line, now featuring an ascending leap rather than a descending one. It also accelerates its rhythm, with two iterations per bar rather than the original one iteration. The key is unstable, switching every two bars between E minor, B♭ major, an ambiguous tonal area (possibly G minor, although this never arrives at its tonic), and A minor. The phrase begins in bar 9 with a repetition of its rhythm, suggesting a standard sentence organization (basic idea, basic idea, fragmentation, fragmentation, cadential idea), but at the location where we would expect a cadence to occur in bars 15–17 the melody instead leaps away from any stable arrival point. The tuning is also distorted. All of these features in combination suggest, both sonically and narratively, that instability predominates, stressing the fact that the protagonist has lost one or more family members. There is a congruence of music with narrative here, once again felt at an emotional level due to the musical structure before the player makes any intertextual associations.

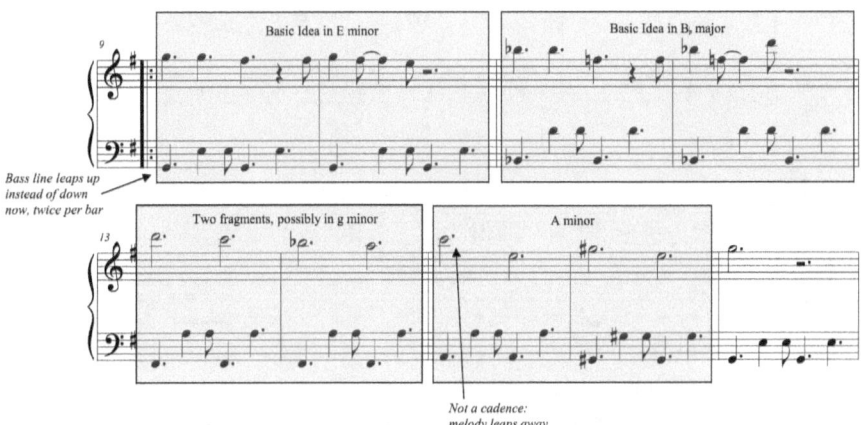

Figure 5.7. Motivic development in the "Death Theme," *Papers, Please*. Source: Composed by Lucas Pope. Reduction and analysis by the author. Courtesy Lucas Pope.

This War of Mine, another indie game focusing on the personal impact of war, takes a very different approach, with a sparse and often melancholic soundtrack. Described as a war survival game, the game is also structured around moral choices and their consequences. The player controls multiple characters trapped in a house during a siege of their city; the gameplay alternates between daytime, where the characters build, repair, and manage their resources, and nighttime, where one character goes out into the city to scavenge resources in dangerous locations. Players must decide whether to steal in order to survive, to trade, or to suffer consequences such as illness or starvation if unable to acquire resources.

Music is omnipresent within the game and reflects this dystopian environment. One song from the game, "Some Place We Called Home" (figure 5.8), features harmonic and tonal instability in a few different ways, and this instability mirrors that of the characters' day-to-day lives.[27] The melody in C minor often resolves to scale-degrees other than the tonic, or in cases where it does resolve to the tonic it is supported by a harmony other than i (see, for example, the resolution to C in bar 3 supported by an A♭ major harmony). Mirroring the narrative elements of unease and conflict even further, the song uses prominent dissonances, with chords sometimes including an added second or fourth (a dissonance against their root, such as the D♭ added against the A♭ major, C minor, and E♭ major chords in bars 6–8) or occurring under a high-register pedal tone, mixing tonic and dominant functions (later in the work and thus not shown in the figure; however, a tonic pedal is audible at 0:50). Digitally altered sounds occur frequently as well, representing the unnatural and

Figure 5.8. Bars 1–12, "Some Place We Called Home," *This War of Mine*. Note the added tones in the basic chord structure and the changes in meter. *Source*: Composed by Piotr Musiał and Krzysztof Lipka. Reduction and analysis by the author. Courtesy 11 bit studios.

uncanny; something does not sound quite right, such as the fade-out of the E♭s in bars 11–12 coupled with the hummed notes in the accompaniment, which include an additional reverb effect. Lastly, a shifting rather than stable meter also unsettles the listener due to some beats lasting longer than expected in the song's 4/4 meter, including the held note in bars 3–6 of the melody. Instead, the repetition of the melody occurs two beats earlier than expected in the established 4/4 meter. While the accompanying gameplay occurs within the safehouse, the music creates unease in the player which then impacts their perception of events to come, a form of evoking.

Music acts to heighten players' emotional engagement in both of these games. Players pick out particular sonic properties that are both inherent musical properties (such as our dissonant suspensions in the example above) as well a cultural topics and tropes (such as slow tempo and minor mode to suggest emotional exhaustion in "Some Place We Called Home") and merge their perception of both dimensions to judge authenticity.

Cheng asks whether such games reveal anything about our true moral selves and suggests that "music, when sounded against (or used to inflict) bodily violence, undergoes ontological violence. It suffers an identity crisis—or, more aptly, becomes a canvas onto which we, its patrons and admirers, project our own crises of identity, epistemology, and humanity."[28] The relationship is two-way. Players map both their their cultural references (in this case of war and conflict) onto their perception of the music, and their understanding of tension and resolution in the music to the game narrative itself. Furthermore, such integration heightens the emotional experience. Cheng, for example, goes on to suggest that a map between musical and emotional instability is not, in fact, unusual, as the discomfort and unease suggested by musical features thereby contribute to the player's emotional engagement with the game. Karen Collins identifies that "sound influences both the functional aspects of gameplay and the emotional connection to the game world."[29] We cannot disentangle perceptions of existential authenticity from constructive authenticity given that they are a single feedback loop.

THE END

And so we come to the end. Video games, as a modern form of media, ultimately communicate in modern ways. In YouTube videos, memes, Twitter posts, or fan sites, we blend pop culture and artistic creation in seamless ways without even thinking about it. And although there is inherently a technological mediation in encountering these forms, in modern society audiences and players are less and less conscious of the device as mediator—they exist on both sides of the magic circle, both inside the game and present as their real selves. As Michel Chion points out,

> sound, much more than the image, can become an insidious means of affective and semantic manipulation. On one hand, sound works on us directly, physiologically (breathing noises in a film can directly affect our own respiration). On the other, sound has an influence on perception: through the phenomenon of added value, it interprets the meaning of the image, and makes us see in the image what we would not otherwise see, or would see differently."[30]

Meaning is fundamental to our understanding of sound in games, and the intersections between the three modes of authenticity suggest that players' mental engagement with the music of games and global digital culture shifts between knowledge, memory, pattern recognition (and its resultant setting up of expectations), and affective response. While game composers and sound designers aim for common tools to set the emotional tone, players engage with their reception of these tools in different ways. Yet nevertheless players manage to create a new, common yet continuously evolving gaming culture. We ultimately cannot unhook our understanding of game music from its larger multimedia and cultural context without doing it a disservice since authenticity is ultimately about linking to this global digital culture and effectively interpreting its cues.

NOTES

1. Note that this has also been a focus of my own previous research; see my analysis of *Assassin's Creed 1* in *The Soundtrack*, which focuses on the correspondences between storyline, shift of gameplay mode, and visuals with sound (Lind, "Music as Temporal Disruption").

2. Alexandra Wilson, "From Authenticity to Anachronism: Pre-Existing Music and 'Epic Englishness' in *Elizabeth* and *Master and Commander*," in *Music in Epic Film: Listening to Spectacle*, ed. Stephen C. Meyer (New York: Routledge, 2017), 110.

3. Ibid., 109.

4. Wilson, "From Authenticity to Anachronism," 120.

5. Ivănescu, *Popular Music*, 24–25.

6. Ibid., 47.

7. Gibbons, *Unlimited Replays*, 33–34. Chion makes a similar observation: media conventions are "determined by a concern for the *rendering* more than for literal truth. We are all thoroughly familiar with these conventions, and they easily override our own experience and substitute for it, become our reference for reality itself" (Audio-Vision, 108).

8. Summers, *Understanding Video Game Music*, 97–101.
9. Hart, "Semiotics in Game Music," 223.
10. Ibid., 221.
11. Bribitzer-Stull, *Understanding the Leitmotif*, 81.
12. *Detroit: Become Human* (Sony Interactive Entertainment, 2018), PlayStation 4 and Microsoft Windows, music by Philip Sheppard, Nima Fakhrara, and John Paesano. The scene can be viewed at https://www.youtube.com/watch?v=tA_NPiMST3E.
13. Katherine Isbister, *How Games Move Us: Emotion by Design* (Cambridge, MA: MIT Press, 2016), 1.
14. Reddit, "Still one of my favorite themes," comment by user rand0me, https://www.reddit.com/r/DetroitBecomeHuman/comments/a83nq8/still_one_of_my_favorite_themes_as_well_as_the/, accessed June 20, 2020.
15. Huron, *Sweet Anticipation*, 292.
16. *Conker's Bad Fur Day* (Rare, 2001), Nintendo 64, music by Robin Beanland.
17. Georges Bizet, *Carmen*, libretto by Henri Meilhac and Ludovic Halévy (Paris: Choudens Père et Fils, 1877).
18. Summers, *Understanding Video Game Music*, 146.
19. For more on this subject, see Tim Summers, "Opera Scenes in Video Games: Hitmen, Divas and Wagner's Werewolves," *Cambridge Opera Journal* 29, no. 3 (2018): 253–286.
20. Cheng, *Sound Play*, 42.
21. Mary Flanagan, *Critical Play: Radical Game Design* (Cambridge, MA: MIT Press, 2009), 243–244.
22. Ibid., 226.
23. Quoted in Flanagan, *Critical Play*, 251.
24. Isbister, *How Games Move Us*, 1.
25. The song is available to hear on Lucas Pope's official YouTube channel at https://www.youtube.com/watch?v=OBQE_TNI7zw.
26. Audio available at https://www.youtube.com/watch?v=J_3Zad-e9f4.
27. *This War of Mine* (11 bit studios, 2014), Microsoft Windows, MacOS X, Linux, Android, iOS, PlayStation 4, Xbox One, and Nintendo Switch, music by Piotr Musiał and Krzysztof Lipka. The soundtrack to the game is available to listen to at https://soundcloud.com/pmcomposer/sets/this-war-of-mine-ost.
28. Cheng, *Sound Play*, 40.
29. Collins, *Playing with Sound*, 56–57.
30. Chion, *Audio-Vision*, 34.

Bibliography

4A Games. *Metro: Exodus*. Deep Silver, 2019. Microsoft Windows, PlayStation 4, Xbox One, Stadia, Luna, PlayStation 5, Xbox Series X/S, Linux, MacOS. Music by Alexei Omelchuk.

11 bit studios and War Child. *This War of Mine*. 11 bit studios, 2014. Microsoft Windows, MacOS X, Linux, Android, iOS, PlayStation 4, Xbox One, and Nintendo Switch. Music by Piotr Musiał and Krzysztof Lipka.

Adams, Ernest. *Fundamentals of Game Design*. 2nd ed. Berkeley, CA: New Riders, 2010.

Apostolopoulos, Hektor. "The Flight Share Thoughts on Their *Assassin's Creed Odyssey* Soundtrack." *Viralbpm*, July 20, 2018. Accessed June 12, 2021. https://viralbpm.com/2018/10/07/the-flight-share-thoughts-on-their-assassins-creed-odyssey-soundtrack/.

Atkinson, Sean. "Soaring Through the Sky: Topics and Tropes in Video Game Music." *Music Theory Online* 25, no. 2 (2019). https://mtosmt.org/issues/mto.19.25.2/mto.19.25.2.atkinson.html.

Bach, Johann Sebastian. *Toccata and Fugue in D Minor*, BWV 565. In *Bach-Gesellschaft Ausgabe*, band 15. Edited by Wilhelm Rust. Leipzig: Breitkopf und Härtel, 1867.

Bethesda Game Studios. *The Elder Scrolls V: Skyrim*. Bethesda Softworks, 2011. Microsoft Windows, PlayStation 3, Xbox 360, PlayStation 4, Xbox One, Nintendo Switch, PlayStation 5, and Xbox Series X/S. Music by Jeremy Soule.

BioWare. *Dragon Age: Inquisition*. Electronic Arts, 2014. Microsoft Windows, PlayStation 3, PlayStation 4, Xbox 360, and Xbox One. Music by Trevor Morris.

———. *Star Wars: Knights of the Old Republic*. LucasArts, 2003. Xbox, Microsoft Windows, MacOS X, iOS, Android, and Nintendo Switch. Music by Jeremy Soule.

Bizet, Georges. *Carmen*. Libretto by Henri Meilhac and Ludovic Halévy. Paris: Choudens Père et Fils, 1877.

Bowden, Sara. "Not Suitable for the Easily Disturbed: Sonic Nonlinearity and Disruptive Horror in *Doki Doki Literature Club*." *The Soundtrack* 11, no. 1 (2020): 7–22.

Braguinski, Nikita. "The Resolution of Sound: Understanding Retro Game Audio Beyond the '8-Bit' Horizon." *NECSUS: European Journal of Media Studies* 7, no. 1 (2018): 105–121.

Bribitzer-Stull, Matthew. *Understanding the Leitmotif: From Wagner to Hollywood Film Music*. Cambridge: Cambridge University Press, 2015.

Brown, Royal S. *Overtones and Undertones: Reading Film Music*. Berkeley: University of California Press, 1994.

Browne, Pierson, and Brian R. Schram. "Intermediating the Everyday: Indie Game Development and the Labour of Co-Working Spaces." In *Game Production Studies*, edited by Olli Sotamaa and Jan Švelch, 83–100. Amsterdam: Amsterdam University Press, 2021.

Buhler, James. "*Star Wars*, Music, and Myth." In *Music and Cinema*, edited by James Buhler, Caryl Flinn, and David Neumeyer, 33–57. Hanover, NH: University Press of New England, 2000.

———. *Theories of the Soundtrack*. Oxford: Oxford University Press, 2018.

Bungie. *Halo: Combat Evolved*. Microsoft Game Studios, 2001. Xbox, Microsoft Windows, MacOS X, and Xbox 360. Music by Martin O'Donnell and Michael Salvatori.

———. *Halo 3*. Microsoft Game Studios, 2007. Xbox 360, Xbox One, Microsoft Windows, and Xbox Series X/S. Music by Martin O'Donnell and Michael Salvatori.

Campbell, Colin. "*Return of the Obra Dinn* Is a Superb Murder Mystery Game." *Polygon* (October 19, 2018). Accessed May 25, 2020. https://www.polygon.com/reviews/2018/10/19/18001242/return-of-the-obra-dinn-review-windows-pc-mac-steam.

Capcom. *Street Fighter II: The World Warrior*. Capcom, 1991. Arcade, Super Nintendo Entertainment System, Amiga, Atari ST, Commodore 64, ZX Spectrum, and PC (DOS). Music by Yoko Shimomura and Isao Abe.

Capcom and Dimps. *Street Fighter IV*. Capcom, 2008. Arcade, PlayStation 3, Xbox 360, Microsoft Windows, iOS, Android, PlayStation 4, and Xbox One. Music by Hideyuki Fukasawa.

Caplin, William. *Classical Form: A Theory of Formal Functions for the Instrumental Music of Haydn, Mozart, and Beethoven*. New York: Oxford University Press, 1998.

Catholic Church. *The Parish Book of Chant*. Richmond, VA: Church Music Association of America, 2008.

Cheng, William. "Monstrous Noise: Silent Hill and the Aesthetic Economies of Fear." In *The Oxford Handbook of Sound and Image in Digital Media*, edited by Carol Vernallis, Amy Herzog, and John Richardson, 173–190. Oxford: Oxford University Press, 2013.

———. *Sound Play: Video Games and the Musical Imagination*. New York: Oxford University Press, 2013.

Chion, Michel. "The Acousmêtre." In *Critical Visions in Film Theory: Classic and Contemporary Readings*, edited by Patricia White, Meta Mazaj, and Timothy Corrigan, 156–165. Boston: Bedford/St. Martin's, 2011.

———. *Audio-Vision: Sound on Screen*. Translated by Claudia Gorbman. New York: Columbia University Press, 1994.

Cole, Yussef. "*Cuphead* and the Racist Spectre of Fleischer Animation." *Unwinnable*, November 10, 2017. https://unwinnable.com/2017/11/10/cuphead-and-the-racist-spectre-of-fleischer-animation/.

Collins, Karen. *Game Sound: An Introduction to the History, Theory, and Practice of Video Game Music and Sound Design*. Cambridge, MA: MIT Press, 2008.

———. *Playing with Sound: A Theory of Interacting with Sound and Music in Video Games*. Cambridge, MA: MIT Press, 2013.

Cook, James. "Game Music and History." In *The Cambridge Companion to Video Game Music*, edited by Melanie Fritsch and Tim Summers, 343–358. Cambridge: Cambridge University Press, 2021.

———. "Playing with the Past in the Imagined Middle Ages: Music and Soundscape in Video Game." *Sounding Out!* (October 3, 2016). https://soundstudiesblog.com/2016/10/03/playing-with-the-past-in-the-imagined-middle-ages-music-and-soundscape-in-video-game/.

Cook, Karen M. "Beyond the Grave: The 'Dies Irae' in Video Game Music." *Sounding Out!* (Blog), December 18, 2017. https://soundstudiesblog.com/2017/12/18/beyond-the-grave-the-dies-irae-in-video-game-music/.

———. "Beyond (the) *Halo*: Chant in Video Games." In *Studies in Medievalism*, edited by Karl Fugelso, vol. XXVII, 183–200. Suffolk, UK: Boydell & Brewer, 2018.

———. "Medievalism and Emotions in Video Game Music." *Postmedieval* 10, no. 4 (2019): 482–497.

———. "Music, History, and Progress in Sid Meier's *Civilization IV*." In *Music in Video Games: Studying Play*, edited by Neil Lerner, K. J. Donnelly, and William Gibbons, 166–182. Routledge Music and Screen Media Series. New York: Routledge, 2014.

———. "'The Things I Do for Lust . . .': Humor and Subversion in *The Bard's Tale*." In *Music in the Role-Playing Game*, edited by William Gibbons and Steven Reale, 21–34. New York: Routledge, 2019.

Creative Assembly. *Medieval II: Total War*. Sega, 2006. Microsoft Windows, MacOS, Linux, Android, and iOS. Music by Jeff van Dyck, Richard Vaughan, and James Vincent.

Dahlhaus, Carl. *Studies on the Origin of Harmonic Tonality*. Translated by Robert O. Gjerdingen. Princeton, NJ: Princeton University Press, 1991. First published 1968 by Bäronreiter.

Dika, Vera. *Recycled Culture in Contemporary Art and Film*. Cambridge: Cambridge University Press, 2003.

Doering, James M. "Status, Standards, and Stereotypes: J. S. Bach's Presence in the Silent Era." *Bach* 50, no. 1 (2019): 5–31.

Ewell, Philip A. "Music Theory and the White Racial Frame." *Music Theory Online* 26, no. 2 (2020). https://doi.org/10.30535/MTO.26.2.4.

Extremely OK Games. *Celeste*. Extremely OK Games, 2018. Linux, MacOS, Microsoft Windows, Nintendo Switch, PlayStation 4, Xbox One, and Stadia. Music by Lena Raine.

Firefly Studios. *Stronghold 3*. 7Sixty, 2011. Microsoft Windows, MacOS X, and Linux. Music by Robert L. Euvino.

Flaherty, Colleen. "Whose Music Theory?" *Inside Higher Ed*, August 7, 2020. https://www.insidehighered.com/news/2020/08/07/music-theory-journal-criticized-symposium-supposed-white-supremacist-theorist.

Flanagan, Mary. *Critical Play: Radical Game Design*. Cambridge, MA: MIT Press, 2009.

Fourcade, Frederic. "The Art of *Shadow of the Colossus* (5/6): Music." *Game Developer*. April 14, 2014. https://www.gamedeveloper.com/audio/the-art-of-shadow-of-the-colossus-5-6-music.
Fox, Toby. *Undertale*. Toby Fox and 8–4, 2015. Microsoft Windows, MacOS X, Linux, PlayStation 4, PlayStation Vita, Nintendo Switch, and Xbox One. Music by Toby Fox.
GameFAQs. "Why is no one talking about the music? *Star Wars Jedi: Fallen Order*." *GameFAQs Message Boards*. Accessed August 22, 2021. https://gamefaqs.gamespot.com/boards/240966-star-wars-jedi-fallen-order/78163722.
Garda, Maria B., and Paweł Grabarczyk. "Is Every Indie Game Independent? Towards the Concept of Independent Game." *Game Studies* 16, no. 1 (October 2016). http://gamestudies.org/1601/articles/gardagrabarczyk.
Gauntlett, Stathis. "Antiquity at the Musical Margins: Rebetika, 'Ancient' and Modern." *Byzantine and Modern Greek Studies* 39, no. 1 (2015): 98–116.
Gibbons, William. *Unlimited Replays: Video Games and Classical Music*. New York: Oxford University Press, 2018.
———. "Wandering Tonalities: Silence, Sound, and Morality in *Shadow of the Colossus*." In *Music in Video Games: Studying Play*, edited by K. J. Donnelly, William Gibbons, and Neil William Lerner, 122–137. New York: Routledge, 2014.
Gilliver, Joe. "Composing Music for Video Games—Key and Tempo." *Game Developer*. Last modified January 10, 2014. https://www.gamedeveloper.com/audio/composing-music-for-video-games---key-tempo.
Grasso, Julianne. "Music in the Time of Video Games: Spelunking *Final Fantasy IV*." In *Music in the Role-Playing Game*, edited by William Gibbons and Steven Reale, 97–116. New York, Routledge, 2019.
Gravois, John. "Knights of the Faculty Lounge." *The Chronicle of Higher Education* 53, no. 44 (July 6, 2007).
Green, Jessica. "Understanding the Score: Film Music Communicating to and Influencing the Audience." *The Journal of Aesthetic Education* 44, no. 4 (2010): 81–94.
Haimo, Ethan. "Developing Variation and Schoenberg's Serial Music." *Music Analysis* 16, no. 3 (1997): 349–365.
Haines, John. *Music in Films on the Middle Ages: Authenticity vs. Fantasy*. Routledge Research in Music 7. New York: Routledge, 2014.
Hart, Iain. "Semiotics in Game Music." In *The Cambridge Companion to Video Game Music*, edited by Melanie Fritsch and Tim Summers, 220–237. Cambridge: Cambridge University Press, 2021.
Hatten, Robert S. *Interpreting Musical Gestures, Topics, and Tropes: Mozart, Beethoven, Schubert*. Bloomington: Indiana University Press, 2004.
———. "The Troping of Topics in Mozart's Instrumental Works." In *The Oxford Handbook of Topic Theory*, edited by Danuta Mirka, 514–538. New York: Oxford University Press, 2014.
Huron, David. "A Comparison of Average Pitch Height and Interval Size in Major- and Minor-Key Themes: Evidence Consistent with Affect-Related Pitch Prosody." *Empirical Musicology Review* 3, no. 2 (2008): 59–63.
———. *Sweet Anticipation: Music and the Psychology of Expectation*. Cambridge, MA: MIT Press, 2006.

The International Arcade Museum. "Pong Doubles." *Museum of the Game*. Accessed August 30, 2021. https://www.arcade-museum.com/game_detail.php?game_id=9075.

Isbister, Katherine. *How Games Move Us: Emotion by Design*. Cambridge, MA: MIT Press, 2016.

Ivănescu, Andra. *Popular Music in the Nostalgia Video Game: The Way It Never Sounded*. Cham, Switzerland: Palgrave Macmillan, 2019.

Japan Studio and Team Ico. *Shadow of the Colossus*. Sony Computer Entertainment, 2005. PlayStation 2. Music by Kow Otani.

Juul, Jesper. *Handmade Pixels: Independent Video Games and the Quest for Authenticity*. Cambridge, MA: MIT Press, 2019.

Kawaguchi, Dais. "*Shadow of the Colossus* Composer Interview, iam8bit 2-LP Vinyl Revealed." *PlayStation.Blog*. Accessed July 6, 2020. https://blog.playstation.com/2018/01/25/shadow-of-the-colossus-composer-interview-iam8bit-2-lp-vinyl-revealed/.

Kolassa, Alexander. "Hail the Nightmare: Music, Sound and Materiality in *Bloodborne*." *The Soundtrack* 11, no. 1 (2020): 23–38.

Korb, Darren. "We Are Supergiant Games, Creators of Hades, Pyre, Transistor, and Bastion. AMA!" *Reddit*. September 22, 2020. https://www.reddit.com/r/NintendoSwitch/comments/ixri6b/we_are_supergiant_games_creators_of_hades_pyre/g68cdou/.

Kuchera, Ben. "It's Time to Take the Animus Out of *Assassin's Creed*: It's the Vegetables of the *Assassin's Creed Universe*." *Polygon*, last updated October 3, 2018. https://www.polygon.com/2018/10/2/17926100/assassins-creed-odyssey-animus.

Laywine, Alison. "*Ancient Greek Music: A Technical History* by Stefan Hagel." *Aestimatio* 9 (2012): 124–170.

Lehman, Frank. "Complete Catalogue of the Themes of Star Wars: A Guide to John Williams's Musical Universe." Last updated February 20, 2022. https://franklehman.com/starwars/.

———. "Manufacturing the Epic Score: Hans Zimmer and the Sounds of Significance." In *Music in Epic Film: Listening to Spectacle*, edited by Stephen C. Meyer, 27–56. New York: Routledge, 2017.

Lerdahl, Fred. *Tonal Pitch Space*. Oxford: Oxford University Press, 2001.

Lind, Stephanie. "Music as Temporal Disruption in *Assassin's Creed*." *The Soundtrack* 11, no. 1 (2020): 57–73.

London, Justin. "Leitmotifs and Musical Reference in the Classical Film Score." In *Music and Cinema*, edited by James Buhler, Caryl Flinn, and David Neumeyer, 85–96. Hanover, NH: University Press of New England, 2000.

Maxis Redwood Shores. *The Sims Medieval*. Electronic Arts, 2011. Microsoft Windows, MacOS X, iOS, and Windows Phone. Music by John Debney.

Medina-Gray, Elizabeth. "Analyzing Modular Smoothness in Video Game Music." *Music Theory Online* 25, no. 3 (2019). https://doi.org/10.30535/MTO.25.3.2.

———. "Meaningful Modular Combinations: Simultaneous Harp and Environmental Music in Two *Legend of Zelda* games." In *Music in Video Games: Studying Play*, edited by K. J. Donnelly, William Gibbons, and Neil Lerner, 104–121. New York: Routledge, 2014.

———. "Modularity in Video Game Music." In *Ludomusicology: Approaches to Video Game Music*, edited by Michiel Kamp, Tim Summers, and Mark Sweeney, 53–72. Sheffield, UK: Equinox Publishing, Ltd., 2016.

———. "Musical Dreams and Nightmares: An Analysis of *Flower*." In *The Routledge Companion to Screen Music and Sound*, edited by Miguel Mera, Ronald Sadoff, and Ben Winters, 562–576. New York: Routledge, 2017.

Mirka, Danuta. "Introduction." In *The Oxford Handbook of Topic Theory*, edited by Danuta Mirka, 1–60. New York: Oxford University Press, 2014.

Mochocki, Michał. "Heritage Sites and Video Games: Questions of Authenticity and Immersion." *Games and Culture* 16, no. 8 (2021): 951–977.

Moore, Allan. "The So-Called 'Flattened Seventh' in Rock." *Popular Music* 14, no. 2 (1995): 185–201.

Munday, Rob. "Music in Video Games." In *Music, Sound and Multimedia: From the Live to the Virtual*, edited by Jamie Sexton, 51–67. Edinburgh: Edinburgh University Press, 2007.

Mundhenke, Florian. "Musical Transformations from Game to Film in *Silent Hill*." In *Music and Game: Perspectives on a Popular Alliance*, edited by Peter Moorman, 107–124. Wiesbaden: Springer Fachmedien Wiesbaden, 2013.

Nakamura, Masato. "Masato Nakamura Interview by *Sonic Central*." By Takashi Iizuka. *Sonic Central*, May 18, 2005. https://web.archive.org/web/20081223022942/http://www2.sega.com/sonic//globalsonic/post_sonicteam.php?article=nakamura.

Neumeyer, David, and James Buhler. *Meaning and Interpretation of Music in Cinema*. Bloomington: Indiana University Press, 2015.

Nintendo EAD. *The Legend of Zelda: Majora's Mask*. Nintendo, 2000. Nintendo 64. Music by Koji Kondo.

———. *The Legend of Zelda: Ocarina of Time*. Nintendo, 1998. Nintendo 64 and GameCube. Music by Koji Kondo.

———. *The Legend of Zelda: Skyward Sword*. Nintendo, 2011. Wii. Music by Hajime Wakai, Shiho Fujii, Mahito Yokota, and Takeshi Hama.

———. *The Legend of Zelda: Twilight Princess*. Nintendo, 2006. Wii and GameCube. Music by Toru Minegishi and Asuka Ohta.

Nintendo EPD. *The Legend of Zelda: Breath of the Wild*. Nintendo, 2017. Nintendo Switch. Music by Manaka Kataoka, Yasuaki Iwata, and Hajime Wakai.

Nordquist, Richard. "The Four Master Tropes in Rhetoric." *ThoughtCo*. Accessed August 11, 2021. https://www.thoughtco.com/master-tropes-rhetoric-1691303.

O'Donnell, Marty. "Bungie's Marty O'Donnell on the *Halo 3* Soundtrack." Interview by Chris Anderson. *Wired* and Chris Anderson, April 24, 2007. https://www.youtube.com/watch?v=b1rMbTLnIbE.

Osborne, Heather. "8-Bit Nostalgia and the Uncanny: Horror as Critique in Twine Games." *Horror Studies* 9, no. 2 (2018): 213–230.

Ordoulidis, Nikos. "The Greek Popular Modes." *British Postgraduate Musicology* 11 (2011): 1–18.

Parker, Felan, and Jennifer Jenson. "Canadian Indie Games between the Global and the Local." *Canadian Journal of Communication* 42, no. 5 (2017): 867–891.

Partch, Harry. *Genesis of a Music: An Account of a Creative Work, Its Roots and Its Fulfillments*. 2nd ed. New York: Da Capo Press, 1974.

Peterson, Richard A. *Creating Country Music: Fabricating Authenticity*. Chicago: University of Chicago Press, 1997.

Plank, Dana. "From the Concert Hall to the Console: Three 8-Bit Translations of the *Toccata and Fugue in D Minor*." *Bach* 50, no. 1 (2019): 32–62.

Plunkett, Luke. "*Assassin's Creed IV*'s Sea Shanties Are a Treasure." *Kotaku*, November 12, 2017. https://kotaku.com/assassins-creed-ivs-sea-shanties-are-a-treasure-1486865100.

Pokorny, Michael, Peter Miskell, and John Sedgwick. "Managing Uncertainty in Creative Industries: Film Sequels and Hollywood's Profitability, 1988–2015." *Competition & Change* 23, no. 1 (2019): 23–46.

Politopoulos, Aris, Angus A. A. Mol, Krijn H. J. Boom, and Csilla E. Ariese. "History Is Our Playground: Action and Authenticity in *Assassin's Creed: Odyssey*." *Advances in Archaeological Practice* 7, no. 3 (2019): 317–323.

Pope, Lucas. *Papers, Please*. Lucas Pope and 3909 LLC, 2013. Microsoft Windows, MacOS X, Linux, iOS, and PlayStation Vita. Music by Lucas Pope.

———. *Return of the Obra Dinn*. Lucas Pope and 3909 LLC, 2018. MacOS, Microsoft Windows, PlayStation 4, Nintendo Switch, and Xbox One. Music by Lucas Pope.

Pugh, Tison, and Angela Jane Weisl. *Medievalisms: Making the Past in the Present*. London: Routledge, 2013.

Quantic Dream. *Detroit: Become Human*. Sony Interactive Entertainment, 2018. PlayStation 4 and Microsoft Windows. Music by Philip Sheppard, Nima Fakhrara, and John Paesano.

Rare. *Conker's Bad Fur Day*. Rare, 2001. Nintendo 64. Music by Robin Beanland.

Reale, Steven. "Analytical Traditions and Game Music: *Super Mario Galaxy* as a Case Study." In *The Cambridge Companion to Video Game Music*, edited by Melanie Fritsch and Tim Summers, 193–219. Cambridge: Cambridge University Press, 2021.

———. "Variations on a Theme by a Rogue A.I.: Music, Gameplay, and Storytelling in *Portal 2*," Parts 1 and 2. *SMT-V* 2, no. 2 (July 2016). http://www.smt-v.org/archives/volume2.html#variations-on-a-theme-by-a-rogue-ai-music-gameplay-and-storytelling-in-portal-2-part-1-of-2.

Respawn Entertainment. *Star Wars Jedi: Fallen Order*. Electronic Arts, 2019. Microsoft Windows, PlayStation 4, Xbox One, Stadia, PlayStation 5, and Xbox Series X/S. Music by Stephen Barton and Gordy Haab.

Roberts, Rebecca. "Fear of the Unknown: Music and Sound Design in Psychological Horror Games." In *Music in Video Games: Studying Play*, edited by K. J. Donnelly, William Gibbons, and Neil Lerner, 138–150. New York: Routledge, 2014.

Roux-Girard, Guillaume. "Listening to Fear: A Study of Sound in Horror Computer Games." In *Game Sound Technology and Player Interaction: Concepts and Developments*, edited by Mark Grimshaw, 192–212. Hershey, PA: Information Science Reference, 2011.

Salen, Katie, and Eric Zimmerman. *Rules of Play: Game Design Fundamentals*. Cambridge, MA: MIT Press, 2003.

Schoenberg, Arnold. *Fundamentals of Musical Composition*, ed. Gerald Strang and Leonard Stein. London: Faber & Faber, Inc., 1967.

Silicon Beach Software. *Dark Castle*. Silicon Beach Software, 1986. Mac, PC, Commodore 64, Amiga, Atari, IIGS, Genesis, CD-I, and MSX. Sound by Eric Zocher.

Sonic Team. *Samba de Amigo*. Sega, 1999. Arcade, Dreamcast, and Wii. Music by Masaru Setsumaru.

———. *Sonic the Hedgehog*. Sega, 1991. Sega Genesis. Music by Masato Nakamura.

Stevens, Richard, and Dave Raybould. "The Reality Paradox: Authenticity, Fidelity and the Real in *Battlefield 4*." *The Soundtrack* 8, nos. 1–2 (2015): 57–75. https://doi.org/10.1386/st.8.1-2.57_1.

Sturges, Robert S. "Medievalism and Periodization in *Frozen River* and *The Second Shepherds' Play*: Environment, Class, Miracle." In *Medieval Afterlives in Popular Culture*, edited by Gail Ashton and Daniel T. Kline, 85–98. New York: Palgrave Macmillan, 2012.

Summers, Tim. "Analysing Video Game Music: Sources, Methods and a Case Study." In *Ludomusicology: Approaches to Video Game Music*, edited by Michiel Kamp, Tim Summers, and Mark Sweeney, 8–31. Sheffield, UK: Equinox Publishing, Ltd., 2016.

———. "Epic Texturing in the First-Person Shooter: The Aesthetics of Video Game Music." *The Soundtrack* 5, no. 2 (2012): 131–151.

———. "Opera Scenes in Video Games: Hitmen, Divas and Wagner's Werewolves." *Cambridge Opera Journal* 29, no. 3 (2018): 253–286.

———. *Understanding Video Game Music*. Cambridge: Cambridge University Press, 2016.

Supergiant Games. *Hades*. Supergiant Games, 2020. MacOS, Microsoft Windows, Nintendo Switch, PlayStation 4, PlayStation 5, Xbox One, and Xbox Series X/S. Music by Darren Korb.

Tenzer, Michael, and John Roeder. *Analytical and Cross-Cultural Studies in World Music*. Oxford: Oxford University Press, 2011.

Traxel, Oliver. "Medieval and Pseudo-Medieval Elements in Computer Role-Playing Games: Use and Interactivity." In *Medievalism in Technology Old and New*, edited by Karl Fugelso and Carol L. Robinson, 125–142. Cambridge: D.S. Brewer, 2008.

Truax, Barry. "Acoustic Space, Community and Virtual Soundscapes." In *The Routledge Companion to Sounding Art*, edited by Marcel Cobussen, Vincent Meelberg, and Barry Truax, 253–263. New York: Routledge, 2017.

TVtropes. "Mary Sue." *TVtropes.org*. Accessed August 4, 2021. https://tvtropes.org/pmwiki/pmwiki.php/Main/MarySue.

———. "Tropes." *TVtropes.org*. Accessed June 8, 2020. https://tvtropes.org/pmwiki/pmwiki.php/Main/Tropes.

———. "Useful Notes/Bollywood." *TVtropes.org*. Accessed June 8, 2020. https://tvtropes.org/pmwiki/pmwiki.php/UsefulNotes/Bollywood.

Ubisoft. "Composing an Epic Score for *Assassin's Creed Odyssey*." *news.ubisoft.com*, October 5, 2018. Accessed July 22, 2020. https://news.ubisoft.com/en-us/article/5Y0vBS44xVo7HJocYvb5Ej/composing-an-epic-score-for-assassins-creed-odyssey.

Ubisoft Montreal. *Assassin's Creed*. Ubisoft Entertainment, 2007. PlayStation 3, Xbox 360, and Microsoft Windows. Music by Jesper Kyd.

Ubisoft Quebec. *Assassin's Creed: Odyssey*. Ubisoft Entertainment, 2018. Microsoft Windows, PlayStation 4, Xbox One, Nintendo Switch, Stadia. Music by The Flight, Mike Georgiades, and Giannis Georgantelis.

Upton, Elizabeth Randall. "Coconut Clops and Motorcycle Fanfare: What Sounds Medieval?" *Sounding Out!* (Blog), September 19, 2016. https://soundstudies blog.com/2016/09/19/coconut-clops-and-motorcycle-fanfare-what-sounds-medieval/.

van Elferen, Isabella. "Analysing Game Musical Immersion: The ALI Model." In *Ludomusicology: Approaches to Video Game Music*, edited by Michiel Kamp, Tim Summers, and Mark Sweeney, 32–52. Sheffield, UK: Equinox Publishing, Ltd., 2016.

———. "Introduction: Sonic Horror." *Horror Studies* 7, no. 2 (2016): 165–172.

———. "Sonic Descents: Musical Dark Play in Survival and Psychological Horror." In *The Dark Side of Game Play: Controversial Issues in Playful Environments*, edited by Torill Elvira Mortensen, Jonas Linderoth, and Ashley ML Brown, 226–241. New York: Routledge, 2015.

———. "¡Un Forastero! Issues of Virtuality and Diegesis in Videogame Music." *Music and the Moving Image* 4, no. 2 (2011): 30–39.

The VGMbassy. Podcast: "Episode 13—The Messenger with Composer Rainbowdragoneyes." Accessed September 5, 2021. https://thevgmbassy.com/2018/10/12/episode-13-the-messenger-with-composer-rainbowdragoneyes/.

West, M. L. *Ancient Greek Music*. Oxford: Clarendon Press, 1992.

Whalen, Zach. "Case Study: Film Music vs. Video-Game Music: The Case of *Silent Hill*." In *Music, Sound and Multimedia: From the Live to the Virtual*, edited by Jamie Sexton, 68–82. Edinburgh: Edinburgh University Press, 2007.

———. "Play Along: An Approach to Videogame Music." *Game Studies* 4, no. 1 (November 2004): paragraph 3. http://www.gamestudies.org/0401/whalen/.

Whalen, Zach, and Laurie N. Taylor. *Playing the Past: History and Nostalgia in Video Games*. Nashville: Vanderbilt University Press, 2008.

Wilson, Alexandra. "From Authenticity to Anachronism: Pre-Existing Music and 'Epic Englishness' in *Elizabeth* and *Master and Commander*." In *Music in Epic Film: Listening to Spectacle*, edited by Stephen C. Meyer, 105–125. New York: Routledge, 2017.

Yee, Thomas B. "Racialized Fantasy: Authenticity, Appropriation, and Stereotype in *Super Mario Odyssey*." Presentation at the North American Conference on Video Game Music, online, June 13, 2021. https://www.youtube.com/watch?v=BKiqMR83410.

Zimmermann, Felix. "Approaching the Authenticities of Late Modernity." In *History in Games: Contingencies of an Authentic Past*, edited by Martin Lorber and Felix Zimmermann, 9–24. Bielefeld: Transcript Verlag, 2020.

Index

8-bit sounds, 127
16-bit sounds, 21, 127
32-bar form, 20

Abe, Isao, 98
accuracy: authenticity compared to, 6; of medievalisms, 71–72, 84–86; Pan-Latinism and, 87–89; in *Samba de Amigo*, for cultural elements, 87–89; Western orientations for, 89. *See also* anachronisms; authenticity
acousmatic sounds, 51–53
"Acre-Underworld" (music track), 76–77, 77
Adams, Ernest, 26
added value principle, in tropes, 63
Aeolian diatonic mode, 17–18, 24n43, 34–36, 83, 93n74; "Song for a Young Girl," 35
Agawu, Kofi, 61
"Age of Oppression/Age of Aggression" (music track), 83–84, *83*
"Agro Falls" (music track), *68*, 69, 157
ALI model, of immersion, 4
"Amen" (music track), 76, 77
anachronisms, in musical medievalisms, 78–79
Analytical and Cross-Cultural Studies in World Music (Roeder), 15
ancient Greek music: in *Assassin's Creed: Odyssey*, 34–40, 39, 48, 56n24; global influences on, 56n33; in *Hades*, 41–43; *Rebetika* folk genre, 31–32; triadic harmonies, 39
animation style, of *Cuphead*, 2
Apostolopolous, Hektor, 40–41
"Ares, God of War" (music track), 35, 36–37
artificial sounds, 21; in music, 12
Assassin's Creed: Odyssey (game), 7, 45; "Acre-Underworld," 76, 77; ancient Greek music in, 38–41, *39*, 48, 56n24; "Ares, God of War," 35, *36*; authenticity of music in, 40–41; "Bacchus Teaches Me to Dance," *37*; "The Black Earth Drinks," *38*; diegetic sounds in, 48; elements of setting in, *29*, 29–30; in Epic game style, 29, 41, 48; establishment of time and place in, 28–41, 43–45; Greek sea shanties in, 16; Ionian diatonic mode in music, 36, *37–38*; language use in, 36; "The Lost Shield," *38*; "Muse of the Forest," *37*; narrative inflection in, 45–48; non-diegetic music in, 48; "Odyssey," 33; as open-world adventure game, 29; Phrygian diatonic mode in music, 35–36, *37–38*; "Poseidon, God of the Sea," *33*, 33–34, *37*, 43; repetition in, 46, 48; silence in, 19; "Song for a Young Girl," 34–35, *35*, *37*; "Song to Bacchus," *37*; "Through the Storm," *38*; "When I Drink," *38*

Assassin's Creed series (game): *Assassin's Creed: Black Flag*, 79; establishment of time and space in, 28–41; Greek sea shanties in, 16; historical authenticity in, 28–30, 86, 154. *See also Assassin's Creed: Odyssey*

associative themes: Bribitzer-Stull on, 7–8; Leitmotif and, 97; thematic representation and, 114–18

Atkinson, Sean, 11, 60

authenticity, 3, 5–10; accuracy compared to, 6; in *Assassin's Creed* series, for historical elements, 28–30, 86, 154; components of, 5–6; constructive, 54, 153, 154–61; definitions of, 5, 153–67; in establishment of time and place, 28–30, 29; existential, 6, 149, 153, 157–67; in indie gaming ethos, 5; magic circle and, 9; media literacy and, 6; model of, 154; of music, in *Assassin's Creed: Odyssey*, 40–41; musical memory and, 6; objective, 6, 54, 82, 149, 153, 154–57; in retro gaming culture, 6; subjective, 6. *See also* anachronisms

"Bacchus Teaches Me to Dance" (music track), 37

Bach, Johann Sebastian, 64–66

"Ballad of the Goddess" (music track), 105

The Bard's Tale (game): folk music in, 94n89; tropes in, 3

basic idea, 20

Battlefield 4 (game), 5

The Battle of Olympus (game), 65

binary beat groupings, 20

Bioshock (game), 137

Bizet, Georges, 159, *160*

"The Black Earth Drinks" (music track), *38*

Bowden, Sarah, 144

Braguinski, Nikita, 8, 125, 127–28, 132–33

Breath of the Wild (game), 101

Bribitzer-Stull, Matthew: on associative themes, 7–8; on Leitmotif, 96–97, 103–4, 118–19

Brown, Eric, 128

Buhler, James, 27, 54n10, 89, 91n16; on Leitmotif, 97; on tropes, 60, 62–64

Burns, Robert, 156

Butler, David, 155

Byzantine system, 38

cadences: harmony and, 18–19; Landini, 84–85; in musical medievalisms, 83–85

Caplin, William, 20, 122n47

Carmen (Bizet), 159, *160*

"Carousel" (music track), *158*

Castlevania (game), 142

Catholicism, plainchants as signifiers for, 74

causal listening, 14

Celeste (game), 20, 100, 126

Celtic culture, in medievalisms, 94n99

character association, thematic representation through, 97–99

Cheng, William, 27, 48–49, 162, 167

Chion, Michel, 14, 63, 168, 168n7

chords: dominant, 18; subdominant, 18; tonic, 18

Christianity. *See* Western European Christianity

cinematic universe, 121n25

Civilization series (game series): *Civilization V*, 89; player-created histories in, 6

Classic Music (Ratner), 61

Collins, Karen, 8, 10, 26–27, 53, 54, 72, 127–28, 141, 149, 167

colonialist themes, 89

Conker's Bad Fur Day (game), 159, 160–61, *161*

consonance, in music, 11–12, 57n47

consonant sounds, 3

constructive authenticity, 54, 153, 155–59; existential authenticity and, 157–59, 161–67

"Continue Theme" (*Sonic the Hedgehog*) (music track), 116–17, *117*

Index

Cook, James: on colonialist themes, 89; on folk music, 78, 82; on historical accuracy, in medievalists tropes, 86–87; as ludomusicologist, 7; on medievalisms, 71, 75–76

Cook, Karen, 6, 60, 69, 108; on folk music, 78, 94n89; as ludomusicologist, 7; on medievalisms, 71–76; on plainchants, 74–77; on *Sacred 2: Fallen Angel*, 74; on tropes, 3, 73, 91n11; on Western orientation for historical accuracy, 89

Critical Play (Flanagan), 162–63

cueing: acousmatic sounds, 51, 53; environmental sounds and, 53; as foreshadowing, 51; as function of sound, 15; in gameworld, 28, 48–53; in *Metro: Exodus*, 49, 50, 51; navigational function and, 52; purpose of, 48; repetition of sounds as, 14; in *Silent Hill* series, 52; in *Skyrim*, 14

Cuphead (game): animation style of, 2; character archetypes in, 3; gender stereotypes in, 2; generational reactions to, 2–3; racial stereotypes in, 2

Dahlhaus, Carl, 84

Dark Castle, 66

"The Dawn Will Come" (music track), 85, *85*

"Death Mountain" (music track), 101

"Death Theme" (music track), 165, *165*

Debney, John, 79, *80–81*

Detroit: Become Human (game), 157–58, *158*, 163

diatonic modes, 17–18; Aeolian, 18, 24n43, 36, 83–84, 93n74; Dorian, 18, 83–85; Ionian, 18, 24n43; Locrian, 18; Lydian, 18, 83; Mixolydian, 18, 83; in musical medievalisms, 83–85; Phrygian, 18, 35–36, 83–84, 93n74

diegetic music: establishment of time and place with, 32–33. *See also* non-diegetic music

diegetic sounds, in *Assassin's Creed: Odyssey*, 48

diegetic sounds, in *Metro: Exodus*, 49

digital distortion, 32

dissonance, in music, 11–12, 57n47, 158

distortion. *See* digital distortion

Doki Doki Literature Club (game), 144–146

dominant chords, 18

Dorian diatonic mode, 18, 83–85

Dragon Age: Inquisition (game), 85, *85*

drones, 36, 82

"Drunken Sailor" (song), 79

Early Music Performance, 5

"Easter eggs," 2

Eco, Umberto, 61

Elder Scrolls V: Skyrim (game), 83–84

"Eldin Volcano" (music track), 102

emotional effects, in player experience, 4

environmental sounds, cueing and, 53

Epic game style: *Assassin's Creed: Odyssey*, 29, 41, 48; definition of, 55n22; musical tropes in, 55n22

ethnomusicology, 15

Euvino, Robert, 75

evoking function, of sound, 15

Evoland 2 (game), 126

Ewell, Philip, 1

existential authenticity, 6, 149; constructive authenticity and, 157–59, 161–67; intertextual referentiality and, 161–62

experience. *See* player experience

fantasy: racialized, 88–89; reality compared to, 23n19, 85–89; in tropes, reality compared to, 85–89

Fantasy genre, 23n19; meaning systems in, 72; open-ended narrative structure of, 72; role-playing in, 70–71; tropes in, 60. *See also* medievalisms

fictive worlds, in gameworld, 26

Final Fantasy (game), 97, 156

final pitch, 56n24

fixed musical scores, 10
Flanagan, Mary, 162–63
The Flight (journal), 30–31, 40–41
focal pitch, 56n24
folk music: in *The Bard's Tale*, 94n89; musical medievalisms and, 78–82; objective authenticity in, 82
"Force Theme" (music track), 106
foreshadowing: cueing as, 51; through Leitmotif, 105–106, 108–13
Fourcade, Frederic, 67
Fox, Toby, 133, 137
framing: as function of sound, 15; musical patterns and, 14
"From Olympus" (music track), 42
Fukusawa, Hideyuki, 98

"Game Over Theme" *(Sonic the Hedgehog)* (music track), 20, 115–17, *116*
gameworld: of *Assassin's Creed: Odyssey*, 25; conceptual approach to, 25–28; cueing in, 28, 48–53; cultural elements of, 26; definition of, 26; establishment of time and place in, 28–45; fictive worlds, 26; game spaces, 26; geography in, 26; history in, 26; immersive experiences in, 27; Leitmotif and, 107–8; magic circle and, 26–27; mythology in, 26; narrative inflection in, creation of, 28; realism in, 53; sound in, construction of, 27; thematic representation in, 100–103; virtuality in, 26. *See also* Fantasy genre; horror genre
gaming culture. *See* gameworld; online gaming
Garda, Maria, 124
Gauntlett, Stathis, 31–32
gender: in *Cuphead*, gender stereotypes in, 2; music theory influenced by, 2
Generation Z, 2
Georgantelis, Giannis, 16, 30, 33–34
Georgiades, Mike, 30
"Ghost Fight" (music track), 136–41, *138*

Gibbons, William, 4–5, 11, 44, 168n7; *The Battle of Olympus*, 65; on historicism in game sounds, 30; on silence as sound choice, 66–67; on tropes, 59, 69–70
Gilliver, Joe, 17
Gorbman, Claudia, 106
"Goron City-Day" (music track), 101, *102*, 104
Grabarczyk, Pawel, 124
Grand Theft Auto (game), 97
Grasso, Julianne, 51
"The Greek Popular Modes" (Ordoulidis), 38
Green, Jessica, 97
Grey, Thomas, 113
Grim Fandango (game), 137
"Guile's Theme" (music track), 98–99

Hades (game): ancient Greek musical influences in, 41–42; "From Olympus," 42; "The House of Hades," 42; "Lament of Orpheus," 42; Locrian diatonic mode in, 42; musical signification in, 41–43; "No Escape," 42–43, 44; "Out of Tartarus," 42; "The Painful Way," 42, 42–43; Phrygian diatonic mode in, 42; "Rage of the Myrmidons," 42
Haimo, Ethan, 111–12
Haines, John, 64, 73, 75; on folk music, 78
Halo series (game), 8, 97, 119; chant melody, *108*, 108–10; *Halo 2*, 109; *Halo 3*, 19, 101, 109, 156; "Keep What You Steal," *113*, 112–13; "Luck," 109–13, *110–11*; "Master Chief," 109
Handmade Pixels (Juul), 5
harmony, 11, 16; in ancient Greek musical scales, 39; cadences and, 18–19
Hart, Iain, 14, 64, 101, 156
Hatten, Robert, 60–61; on topic theory, 63–64, 70
Henson, Joe, 30

horror genre, in videogames: clichés in, 51; *Metro: Exodus* (game), 49, 51; navigational function and, 52; non-diegetic sounds in, 49; *Silent Hill*, 27, 49; tropes in, 64–65
"The House of Hades" (music track), 42
Huron, David, 17, 20, 114–15, 118, 159

idées fixe, Leitmotif and, 97
immersive experience: ALI model, 4; in gameworld, 27
incongruence, of music, 10
indie gaming: authenticity and, 5. *See also specific games*
in-game triggers, for music, 10
instrumentation, 10, 21
intertextuality: existential authenticity and, 161–62; perceptions of, 21; player experience and, 4; tropes and, 60
Ionian diatonic mode, 18, 24n43; in *Assassin's Creed: Odyssey* music, 35–36, *37–38*
Isbister, Katherine, 158, 163
Ivănescu, Andra, 8, 137, 142; on nostalgia games, 140, 148; on retro games, 125–26, 133, 140

Jackson, Roman, 61
jazz music: as musical influence, in video games, 20–21; in retro games, 137–39
Jenson, Jennifer, 124
"John Barleycorn" (song), 78–79, 156
"Johnny Boker" (song), 79
Jones, Quincy, 89
Juul, Jesper, 5

"Keep What You Steal" (music track), *113*, 112–13
Kiary Games, 3
Kline, Daniel, 71–72
Kolassa, Alexander, 48
Korb, Darren, 41–43
Kotaku, 79
Kyd, Jesper, 76, 77

"Lament of Orpheus" (music track), 42
Landini cadences, 85
language use: in *Assassin's Creed: Odyssey*, 36; in tropes, 62
LA Noire (game), 137, 155
layering, of music, 10
The Legend of Zelda series (game), 8, 69, 97, 142; "Ballad of the Goddess," 105; "Death Mountain," 101, *102*; "Goron City-Day," 101, *102*, 104; *The Legend of Zelda: Breath of the Wild*, 101; *The Legend of Zelda: Link's Awakening*, 125; *The Legend of Zelda: Majora's Mask*, 101, 104; *The Legend of Zelda: Ocarina of Time*, 100–103, 104–5; *The Legend of Zelda: Skyward Sword*, 103, 105; *The Legend of Zelda: Twilight Princess*, 101–2; thematic representation in, 100–103; Whalen analysis of, 100; "Zelda's Lullaby," 105
Lehman, Frank, 55n22, 106
Leitmotif, 103–7; alterations and modifications of, 97; associative themes and, 97; authenticity of gameworld and, 107–8; Bribitzer-Stull on, 96–97, 103–4, 118–19; Buhler on, 97; components of, 103–4; conceptual approach to, 7–8, 95–97; emotional elements of, 104; foreshadowing through, 108–13; function and purpose of, 96–97; idées fixes and, 97; motto themes and, 97; musical representation and, 7–8; musical symbolism and, 97; mythic effects and, 106; narrative elements as parallel to, 96–97
Lerdahl, Fred, 105
Limbo (game), 58n57
Lipka, Krzysztof, *166*, 166–67
listening. *See* causal listening; semantic listening
liturgical plainchants. *See* plainchants
Locrian diatonic mode, 18; in *Hades* music, 42
London, Justin, 104

looping, of music, 10, 15; melodic, 20; repetition and, 19–20
Lord of the Rings series (Tolkien), 70
"The Lost Shield" (music track), *38*
"Luck" (music track), 109–13, *110*
ludomusicology, 7
Lydian diatonic mode, 18, 83
Lyraki, Kalia, 30

magic circle, 9, 26; tropes and, 59
major scales, 17–18, 24n43
Mamoulian, Rouben, 65
The Mandalorian (TV series), 106
maqam system, 38
Martin, Ricky, 89
massive multiplayer online role-playing games, 70
McLuhan, Marshall, 163
Meaning and Interpretation of Music in Cinema (Buhler and Neumeyer), 91n16
media literacy, authenticity and, 5
mediation, player experience and, 25
media tropes, 64, 89–90
Medieval II: Total War (game), 76, 77
medievalisms: Celtic culture in, 94n99; Cook, James, on, 71, 75–76; Cook, Karen, on, 71–77; creative interpretations of, 71–72; definitions of, 71–73; in Fantasy genre, 70–72; historical accuracy of, 72, 85–87; modern research on, 71; musical, 71–85; Nordic culture in, 94n99; plainchants, 73–77, 77; tropes and, 70–85
Medina-Gray, Elizabeth, 10, 57n47
melodic looping, 20
melody, 11, 18
memory. *See* musical memory
The Messenger (game), 20, 126, 128
Metro: Exodus (game): cueing in, 49, 50, 51, 156–57; player-generated sounds, 49, 51–52; soundscape analysis, 50
minor scales, 17–18, 24n43
Mirka, Danuta, 61
Mixolydian diatonic mode, 18, 83

"Moldheart's Hornpipe" (music track), 79, *80–81*, 84
Monelle, Raymond, 61
Moore, Allan, 131
motto themes, 97
Munday, Rod, 97, 106
Mundhenke, Florian, 49, 53
"Muse of the Forest" (music track), *37*
music, in video games: analysis of, 10–21; artificial sounds in, 12; congruence of, 10; consonance, 11–12; disjunction of, 10; dissonance, 11–12; drones in, 36; van Elferen on, 13; form of, 16; harmony, 11; hierarchic perceptions of, 20; incongruence of, 10; in-game triggers for, 10; instrumentation, 10, 21; jazz influences on, 20–21; layering of, 10; looping, 10, 15; melody, 11; natural sounds in, 12; patterns in, 13–14; phrasing, 16; player satisfaction from, 13; popular music as influence on, 10; in retro games, 127–33; rhythm, 10; shifting temporality of, 13; smoothness of, 10; tonality, 10–11. *See also* non-diagetic music; *specific topics*
musical medievalisms, 71–85; anachronisms in, 79; cadence in, 83–85; diatonic modes in, 83–85; in Fantasy games, 7; in folk music, 78–82
musical memory, authenticity and, 6
musical structure, gameplay experience influenced by, 3–4
musical symbolism, Leitmotif and, 97
musical tropes: in Epic game style, 55n22; in medievalisms, 73; *Toccata and Fugue in D minor*, 64–66, *65*
music theory: conservatism in, 1; contemporary, 1; foundations in, 17; gender as influence on, 2; generational influences on, 2; purpose of, 1; transcriptions in, 17; tropes and, 60; white racial frame, 1. *See also specific topics*

"My Best Friend" (music track), 147
mythic effects, Leitmotif and, 106

Nakamura, Masato, 115–16, 122n46
narrative inflection: in *Assassin's Creed: Odyssey*, 45–48; in gameworld, 28
natural sounds, 21; in music, 12
Neumeyer, David, 27, 54n10, 91n16; on tropes, 60, 62–64
Ninja Gaiden (game), 128
Nitsche, Michael, 26
"No Escape" (music track), 42–43, 44
non-diegetic music: in *Assassin's Creed: Odyssey*, 48; establishment of time and place through, 32
non-diegetic sounds, in *Metro: Exodus*, 49
non-linear sounds, 19
non-player characters, 33
Nordic culture, in medievalisms, 94n99
North American Conference on Video Game Music, 60
nostalgia: as game theme, 2; Ivănescu on, 125
nostalgia games, retro games as distinct from, 140, 148

objective authenticity, 6, 54, 149, 153, 153–57; in folk music, 82
octachords, 39
O'Donnell, Martin, 108, 108–10, 110
"Odyssey" (music track), 33
online gaming, 90
open-world adventure games, *Assassin's Creed: Odyssey*, 29
Ordoulidis, Nikos, 38–39
Orestes (Euripides), 36
Osborne, Heather, 148
Otani, Kow, 69
out-of-game knowledge, of players, 4
"Out of Tartarus" (music track), 42
"Out of the Cold" (music track), 84, 84

"The Painful Way" (music track), 42, 42–43
"A Pane in the Glass" (music track), 75–76, 76

Pan-Latinism, 88–89
Papers, Please (game), 126–27, 163–65; "Death Theme," 165, 165; "Glory to Arztotzka," 164–65, 164–65
Parker, Felan, 124
Partch, Harry, 56n33
patterns, musical, 13–14; framing and, 14; in *Super Mario*, 14; virtual geographies influenced by, 14
Peterson, Richard, 5
phrasing, musical, 16
Phrygian diatonic mode, 18, 83–84, 93n74; "Ares, God of War," 35; in *Assassin's Creed: Odyssey* music, 35–36, 37–38, 56; in *Hades* music, 42; "Song for a Young Girl," 35
Pierce, Charles Sanders, 61
pitch center, 56n24
place. *See* time and place
plainchants, liturgical, 73–77; blending of modern musical elements in, 77; as Catholic signifier, 74; Cook, Karen, on, 74; fragments of, 76–77, 77
Plank, Dana, 65–66
player experience: emotional effect, 4; expectations for gameworld as factor in, 25; intertextuality and, 4; media literacy and, 4; mediation as part of, 25; from musical elements, player satisfaction from, 13; musical structure as influence on, 3–4; out-of-game knowledge as influence on, 4; physical interaction as element of, 4; sonic structure as influence on, 3–4; tonality as influence on, 17–18. *See also* authenticity; immersive experience
player-generated sounds, in *Metro: Exodus*, 49, 51–52
players: creation of own histories, 6–7; out-of-game knowledge of, 4
Playing the Past (Taylor and Whalen), 25, 59
Playing with Sound (Collins), 26
Plunkett, Luke, 79
Pokémon (game series), 97

188 Index

Pong (game), 97
Pope, Lucas, 126–27, *129–30*, 163, 164–65, *164–65*
popular music: establishment of time and place with, 32; music in video games influenced by, 20–21
Popular Music in the Nostalgia Video Game (Ivănescu), 137–38
Portal (game), 3
"Poseidon, God of the Sea" (music track), *33*, 33–34, 37, 43
Pugh, Tison, 26, 71, 72

race: in *Cuphead*, racial stereotypes in, 2; music theory informed by, 1
racialized fantasy, 88–89; colonialism and, 89
"Rage of the Myrmidons" (music track), 42
Ratner, Leonard, 61
Reale, Steven, 11
realism: in gameworld, 53; in tropes, fantasy compared to, 85–89. *See also* authenticity
reality, in tropes, fantasy compared to, 85–89
Rebetika folk genre, 32
repetition: in *Assassin's Creed: Odyssey*, 46, 48; as cueing, 14; form structures through, 19–21; looping and, 19–20; in *Sonic the Hedgehog*, 20
"Resurrection" (music track), 67, *68*, 96
retro games: acoustic sounds in, 128; *Celeste*, 20, 100, 126; conceptual approaches to, 123; cultural/historical referencing in, 136–40; digital sounds in, 128; evocation of retro themes, 126–33; existential authenticity in, 149; Ivănescu on, 125–26, 133, 140; jazz music in, 137–39; *The Messenger*, 20, 126; music features in, 127–33, *128*; *Ninja Gaiden*, 128; nostalgia and, 125–26; nostalgia games as distinct from, 140; *Papers, Please*, 126–27, 163; pastiche in, 126; *Return of the Obra Dinn*, 8, 126–27, *129–30*, 129–33, *132*, 149, 163; 16-bit sounds in, 21; superimposing in, 126; *Undertale*, 8, 16, 100–101, 126, 133–47; Unity design tool and, 124
retro gaming culture: authenticity in, 6; Greenlight program and, 124; indie gaming and, 124; nostalgia and, 125–26; in twenty-first century, expansion of, 123–24; Valve Steam service and, 124
Return of the Obra Dinn (game), 8, 126–27, 131, 148, 161; "Soldiers of the Sea," *129–30*, 129–30, *132*, 132–33
rhythm, 10
rhythm, transcriptions and, 16
Roberts, Rebecca, 48
Roeder, John, 15
Rohan, Emma, 30
role-playing: in Fantasy genre, 70–71; massive multiplayer online role-playing games, 70
Roux-Girard, Guillaume, 48
"Ruins" (music track), 134–36, *135*, *137*, 141
Rules of Play (Salen and Zimmerman), 26

sacred, as concept, 92n37
Sacred 2: Fallen Angel (game), 74–75
Salen, Katie, 26
Samba de Amigo (game), 7, *87*, 155; cultural accuracy in, 87–88; Pan-Latinism in, 88–89
Saussure, Ferdinand de, 61
scale, 18
schemas, tropes and, 59–60
Selling, Kim, 78
semantic listening, 14
semiotics, tropes and, 61
Shadow of the Colossus (game), 157–58; academic scholarship on, 68; "Agro Falls," *68*, 69, 158; "Resurrection," 67, *68*, 96; silence as sound choice in, 66–67; tropes in, 66–70
Shimomura, Yoko, 98

signifiers: in establishment of time and place, 30; plainchants as, for Catholicism, 74
silence, as sound choice: in *Assassin's Creed: Odyssey*, 19; Gibbons on, 66–67; in *Shadow of the Colossus*, 66–67
Silent Hill series (game), 27, 49; cueing in, 52
Sims Medieval (game), 79–82
Skyrim series (game): *Elder Scrolls V: Skyrim*, 83–85; musical cues in, 14; "Out of the Cold," 84–5, *84*
"Sloprano" (music track), 159, *160*, 161
Smith, Alexis, 30
"Soldiers of the Sea" (music track), 127, *129*, 129–33; chord mapping in, *132*, 132–33
"Some Place We Called Home" (music track), *166*, 166–67
"Song for a Young Girl" (music track), 34–35, *35*, *37*; Aeolian diatonic mode, 35; Phrygian diatonic mode, 35
"Song to Bacchus" (music track), *37*
sonic structure, gameplay experience influenced by, 3–4
Sonic the Hedgehog (game), 8, 114; "Continue Theme," 116–17, *117*; "Game Over Theme," 20, 115–17, *116*; repetition in, 20; "Title Theme," 113, *115*, 117
Sound Play (Cheng), 49, 162
sounds, in video games: artificiality of, 21; cueing, 15; evoking function, 15; framing function, 15; functions of, 15, 27–28; historicism in, 30–32; meaning in, 166; non-linear, 19; texturing function, 15. See also diegetic sounds; music
Star Wars series (game), 8; "Force Theme," 106–7; *Rogue One: A Star Wars Story* (film), 106; *Star Wars Jedi: Fallen Order*, 106; *Star Wars: Knights of the Old Republic*, 107
Stefos, Panagiotis, 30
stereotypes, *Cuphead*, 2

Street Fighter series (game), 8; "Guile's Theme," 98–99; *Street Fighter II*, 98–99; *Street Fighter IV*, 98–99
Stronghold 3 (game): "John Barleycorn," 78–79, 156; "A Pane in the Glass," 75–76, *76*; "Tom of Bedlam," 78–79
Studies in the Origin of Harmonic Tonality (Dahlhaus), 84
Sturges, Robert S., 71
subdominant chords, 18
subject authenticity, 6
Summers, Tim, 4–5, 40, 44–45, 60, 121n32; on fixed musical scores, 10
Super Mario series (game), 97, 154; musical patterns in, 14; *Super Mario Odyssey*, 88
Super Smash Bros. Ultimate (game), 99, 156
Sweeney, J. W., 79
Sweet Anticipation (Huron), 114–15

"Take on Me" (music track), 89
Taylor, Laurie, 25, 53–54, 59
"Temmie Village" (music track), 100–101
Tenzer, Michael, 15
texturing function, of sound, *15*
thematic representation: advanced applications of, 107–13; associative themes, 114–18; through character association, 97–99; conceptual approach to, 96–97; configurative signs in, 101; through gameworld association, 100–102; integration of themes, 118–19; in *The Legend of Zelda* series, 100–102. See also Leitmotif
This War of Mine (game), 166–67; "Some Place We Called Home," *166*, 166–67
"Through the Storm" (music track), *38*
timbre, 21
time and place, in gameworld, establishment of, 28–45; analysis of, 32–39; in *Assassin's Creed* series, 28–41, 43–45; diachronic

Greek culture and, 32; diegetic music, 32–33; digital distortion in, 32; focus on feel for, 31; historical authenticity in, 28–30, *29*; historicism in game sounds, 30–32; non-diegetic music, 32; non-player characters, 33; popular music styles in, use of, 32; *Rebetika* folk genre and, 32; signifiers in, 30

Tiny Room Stories (game), 3

"Title Theme" (*Sonic the Hedgehog*) (music track), 113, *115*, 117

Toccata and Fugue in D minor (Bach), 64–66

Tolkien, J. R. R., 70

"Tom of Bedlam" (song), 78–79

tonality, of music: assessment of, 57n47; player perception influenced by, 17–18

tonic chords, 18

tonic pitch, 56n24

topic theory, classical: Hatten on, 63–64, 70; tropes and, 61, 63–64

transcriptions, 14–17; in music theory, 17; pitch and, 16; rhythm and, 16; volume and, 16

Traxel, Oliver, 71–72, 85–86

tropes: added value principle and, 63; in *The Bard's Tale*, 3; Buhler on, 60, 62–64; classical topic theory, 61, 63–64; conceptual approach to, 59–60; Cook, Karen, on, 3; as cultural shorthand, 59; definitions of, 61–64; in Fantasy genre, 60; fluidity of, 62–63; Gibbons on, 59, 69–70; in horror genre, 64–65; intertextuality and, 60; language use in, 62; magical circle and, 59; media, 64, 89–90; medievalisms and, 70–85; musical, 55n22, 64–66, *65*, 73; musical quotation as, 64–66; in music scholarship, 61; music theory and, 60; Neumeyer on, 60, 62–64; player choice and, 63–64; reality in, fantasy compared to, 85–89; schemas and, 59–60; semiotics and, 61; in *Shadow of the Colossus*, 66–70; theoretical process as, 66–70; *Toccata and Fugue in D minor*, 64–66, *65*

Truax, Barry, 141

"True Pacifist Route," 142

"Tubthumping" (music track), 89

TVTropes.com, 61–62

Understanding Video Game Music, 40

Undertale (game), 8, 16, 133–47; "Ghost Fight," 136–41, *138*; "My Best Friend," 146–47; Napstablook character in, 139–40, *139–40*; "Ruins," 134–36, *136–37*, 141; soundscape analysis of, narrative interactions in, 141–47, *144*; "Temmie Village," 100–101; visual styles in, *146*, 146–47; "Your Best Friend," 142–143, *143*

Unlimited Replays (Gibbons), 4, 11, 30, 168n7

Upton, Elizabeth Randall, 5, 78

Valve Studios, 3

Van Dyck, Jeff, 76

van Elferen, Isabella, 48; ALI model of immersion, 4; on gameworld, 26–27; on magic circle, 9; on musical role in video games, 13; on role of game soundtrack, 53; on virtuality, 26

video games: analytical approach to, 10–21; as cultural construction, 1–4; "Easter eggs," 2

virtuality, 9–10; van Elferen on, 26; in gameworld, 26; magic circle, 9

Wagner, Richard, 104

Wang, Nina, 6

Weisl, Angela, 26, 71, 72

West, M. L., 34, 36, 56n24
Western European Christianity, 92n37
Whalen, Zach, 25, 27, 48–49, 53–54, 59; on *The Legend of Zelda,* analysis of, 100
"When I Drink" (music track), *38*
Williams, John, 106–7

Yee, Thomas, 88
"Your Best Friend" (music track), *143*

"Zelda's Lullaby" (music track), 105
Zimmer, Hans, 55n22
Zimmerman, Eric, 26
Zimmerman, Felix, 6

About the Author

Stephanie Lind is associate professor at the Dan School of Drama and Music, Queen's University, where she teaches courses in music theory and musicology. With interests in the music of video games, Canadian music, and post-tonal theory, her perspective on music analysis focuses on listening, perception, and audience reception as well as the interaction of sound and other components within mixed media forms. She has written for journals including *Intersections, Perspectives of New Music,* and *Music Theory Online,* and has presented work in ludomusicology at the North American Conference for Video Games, Congress, and the Society for Music Theory. Her written scholarship includes a guest-edited issue of *The Soundtrack* on musical disruption in video games, a chapter in *Music Video Games: Performance, Politics, and Play,* and several upcoming book chapters.

www.ingramcontent.com/pod-product-compliance
Lightning Source LLC
Chambersburg PA
CBHW020120010526
44115CB00008B/903